Principles of Basic Construction Economics in the 21st Century

Principles of Basic Construction Economics in the 21st Century

Andrew Ebekozien and
Clinton Aigbavboa

Published by Emerald Publishing Limited, Floor 5, Northspring, 21-23 Wellington Street, Leeds LS1 4DL.

ICE Publishing is an imprint of Emerald Publishing Limited

Other ICE Publishing titles:
Financing Infrastructure Projects: A practical guide, Second edition
Tony Mema and Faisal F. Al-Thani. ISBN 9780727763365
Procurement and Contract Strategies for Construction
Ian Heaphy. ISBN 9780727763716
Civil Engineering Procedure, Eighth edition
The Institution of Civil Engineers. ISBN 9780727764270

A catalogue record for this book is available from the British Library

ISBN 978-1-83549-841-5

© 2024 Andrew Ebekozien and Clinton Aigbavboa. Published under exclusive licence by Emerald Publishing Limited.

Permission to use the ICE Publishing logo and ICE name is granted under licence to Emerald from the Institution of Civil Engineers. The Institution of Civil Engineers has not approved or endorsed any of the content herein.

All rights, including translation, reserved. Except as permitted by the Copyright, Designs and Patents Act 1988, no part of this publication may be reproduced, stored in a retrieval system or transmitted in any form or by any means, electronic, mechanical, photocopying or otherwise, without the prior written permission of the publisher, Emerald Publishing Limited, Floor 5, Northspring, 21-23 Wellington Street, Leeds LS1 4DL.

This book is published on the understanding that the author is solely responsible for the statements made and opinions expressed in it and that its publication does not necessarily imply that such statements and/or opinions are or reflect the views or opinions of the publisher. While every effort has been made to ensure that the statements made and the opinions expressed in this publication provide a safe and accurate guide, no liability or responsibility can be accepted in this respect by the author or publisher.

While every reasonable effort has been undertaken by the author and the publisher to acknowledge copyright on material reproduced, if there has been an oversight please contact the publisher and we will endeavour to correct this upon a reprint.

Cover photo: Lucian Coman/Shutterstock

Commissioning Editor: Michael Fenton
Content Development Editor: Cathy Sellars
Books Production Lead: Emma Sudderick

Typeset by KnowledgeWorks Global Ltd.
Index created by David Gaskell

Contents

About the authors   vii

01   **The construction industry and economics**   1
- 1.1. Nature and scope of activities in the construction industry   1
- 1.2. Characteristics of the construction industry   3
- 1.3. Major stakeholders in the building industry   4
- 1.4. The construction industry as an economic regulator   8
- 1.5. Effects of government action on the construction industry   9
- 1.6. The history of building economics   10
- 1.7. Construction economics and its scope   11
- 1.8. Summary   12

02   **Approximate estimating and design variables in the construction industry**   15
- 2.1. Approximate estimating   15
- 2.2. Forms of approximate estimating   15
- 2.3. Design variables   21
- 2.4. Summary   25

03   **Cost planning, control and analysis of construction projects**   27
- 3.1. Design team's role in contract administration   27
- 3.2. Cost planning   36
- 3.3. Cost planning techniques   38
- 3.4. Sources of cost information   39
- 3.5. Cost control   40
- 3.6. Definition of terms   40
- 3.7. Cost analysis   41
- 3.8. Summary   48

04   **Life-cycle costing for building projects**   51
- 4.1. Introduction to life-cycle costing   51
- 4.2. Fundamentals of life-cycle costing   51
- 4.3. Conducting a life-cycle cost analysis   52
- 4.4. Encumbrances to the application of life-cycle costing in construction   53
- 4.5 Life-cycle costing methods   55
- 4.6. Tools and techniques for life-cycle costing (investment appraisal)   55
- 4.7. Summary   65

05	**Cost indices in the construction industry**	**67**
	5.1. Introduction	67
	5.2. Relevance of cost and price indices in a developing country's construction industry	67
	5.3. Types of cost indices	68
	5.4. The hedonic regression method	72
	5.5. Challenges in choosing cost indices	72
	5.6. Summary	73
06	**Property valuation and developer's budget**	**75**
	6.1. Methods of determining the value of property	75
	6.2. Investment	75
	6.3. Methods of valuation	77
	6.4. Developer's budget	79
	6.5. Summary	88
07	**Economics of sustainable construction**	**91**
	7.1. Introduction to sustainable construction	91
	7.2. Benefits of sustainable construction to the industry	92
	7.3. Principles of sustainable construction	93
	7.4. Encumbrances to economic sustainable construction	94
	7.5. Encumbrances of sustainable construction in developing countries	95
	7.6. The way forward	96
	7.7. Summary	98
08	**Economics of smart construction**	**103**
	8.1. Introduction and foundations of smart construction	103
	8.2. Benefits of smart construction to the industry	105
	8.3. Encumbrances facing smart construction	106
	8.4. The future of smart construction	108
	8.5. Summary	110
	Index	**115**

About the authors

Dr Andrew Ebekozien is a Senior Research Associate at the Department of Construction Management and Quantity Surveying, University of Johannesburg, South Africa. He is also a Lecturer at Auchi Polytechnic, Auchi, Nigeria.

Professor Clinton Aigbavboa is the Director of the NRF/DSI Research Chair in Sustainable Construction Management and Leadership in the Built Environment and of the cidb Centre of Excellence at the University of Johannesburg, Johannesburg, South Africa.

Principles of Basic Construction Economics in the 21st Century
Andrew Ebekozien and Clinton Aigbavboa
ISBN 978-1-83549-841-5
https://doi.org/10.1108/978-1-83549-838-520241001
Emerald Publishing Limited: All rights reserved

Chapter 1
The construction industry and economics

1.1. Nature and scope of activities in the construction industry

Globally, the construction industry has no universal definition, yet it is one of the most critical industries in the socio-economic development of developed and developing countries. The sector can influence most economic sectors (finance, industry and commerce), contributing significantly to infrastructure advancement and gross domestic product (GDP). It has been argued that the construction industry in fact comprises many industries (Ofori, 2015) and it has also been reported that the industry accounts for about 6% of global GDP (WEF, 2018). According to Ofori (2012), the sector contributes 3–10% of GDP and creates job opportunities for about 10% of a nation's employees. In developing countries (e.g. Nigeria, Ghana and South Africa), the GDP contribution is low compared with that in developed countries (e.g. Singapore, the UK and Korea). Understanding the nature, important attributes and the requirements of the construction industry – also known as the 'built environment sector' – is pertinent. The sector is important because it adds to personal satisfaction through engineering and services provided to end users (Ebekozien, 2022). Thus, it is a critical sector in every economy and one of the most significant economic industries across the world (Delgado et al., 2019). However, despite its economic importance, the industry is beset with inefficiencies, especially in developing countries.

The Nigerian built environment sector is a remarkable, complex and regularly divided industry. In its sectoral classification of the economy, it is also known as the building and construction sector. The built environment sector comprises a wide range of inexactly incorporated associations that builds, modifies and fixes a wide scope of various building structures, along with structural designing and heavy engineering work. The boundaries between these are obscured, but incorporate planning, guidelines, production, installation and the upkeep of structures. The sector has a duty regarding infrastructure advancement. In Ghana, the sector is the second largest GDP contributor (13.7%) and employs about 7% of the working population (Boadu et al., 2020).

The industry is multi-faceted, with evidence of encumbrances, especially in developing countries. Nigeria's construction sector faces impediments to project performance, which has resulted in a decline of the national GDP (Unegbu et al., 2022). Olanrewaju et al. (2020) found that Nigeria has enhanced non-collaborative work practices in some construction projects. Developing countries, Nigeria included, have been identified as fertile ground for the construction industry (Akdag and Maqsood, 2019). However, some studies (e.g. Ebekozien, 2020; Nwachukwu and Nzotta, 2010) have noted the poor performance of construction projects (roads, bridges, buildings, dams etc.) in many developing countries. This can be traced to several factors, including corruption and lax utilisation of project management best practices. In Ghana, several small organisations characterise the industry (Boadu *et al.*, 2020), possibly because of relaxed entry requirements. In other words, business entities and individuals without the basic requirements and personnel can register as construction organisations (Ofori-Kuragu, 2013). The industry accumulates little capital compared with the manufacturing and processing industries. The built environment sector could be regarded as an assembly sector, coupling the finished goods of other sectors on site. Contract building plans and other relevant documents depict the architect's original goals, and trained workers are guided by professional supervisors designing, constructing and coupling various construction parts on site. Division 45 of the Revised 2003 (as cited in Winch (2003)) defined the construction industry by Standard Classification as follows.

- General building and civil engineering construction.
- Construction and repair of building projects – the developing, upgrading and renovating of structures. This includes specialists engaged in segments of development and maintenance work such as block-laying and the erection of steel and concrete works.
- Civil engineering: construction of pipelines, tunnels, runways, roads, railways, airports and so on.
- Installation of fixtures and fittings (plumbing, electrical fixtures etc.).
- Building completion – for example joinery, plastering, painting, decorating and so on.
- Heavy engineering work: construction of power generating plants, turbines, petrochemical factories, gas plants, refineries, farm tanks, cement plants, sugar plants, shipyards, aluminium plants and the like.

In advanced countries such as Singapore, the UK and Korea, the construction industry is the largest in terms of job creation. In developing countries such as Ghana and Nigeria, the agricultural sector ought to be the largest employer, followed by the construction sector. However, this is not the case today in Nigeria. The construction industry is thus critical in the drive for the economic development of developing nations. Construction is a global task centred on infrastructural and industrial advancement (Asare *et al.*, 2022). In 2013, the World Bank reported that

- many Nigerians lack access to basic public services (electricity, roads, pipe-borne water etc.)
- where they are available in cities, the facilities need to be improved (World Bank, 2013).

The industry cannot impact the demand for outcomes or regulate the supply. This implies that several factors (social, economic and environmental) influence the degree of activity in the sector. The need for construction projects is influenced by several variables, as reported by Ebekozien and Aigbavboa (2024).

- The built environment is susceptible to economic impacts, as witnessed by the world economic downturn in 2008. This is possibly one of the reasons why governments

regularly used the construction sector as a mechanism to regulate the economy. This can take the form of fluctuating interest rates to control the demand for residential buildings.
- The public sector is the client/employer of about 50% of construction works. This makes it easy to cut public sector spending on construction works such as roads, hospitals, school buildings and so on.
- Demand can come from a variety of sources – from the construction of one-bedroom apartments to megaprojects, such as World Cup tournaments.
- An optimistic construction market needs accessibility to realistic cost credit.
- State and federal governments can influence the demand for construction by allowing tax breaks for some projects.
- The repair and maintenance section of the sector's output covers about 50%, and this can be negatively affected in times of economic downturns.

1.2. Characteristics of the construction industry

The construction industry engages in the production and operation of buildings and civil engineering works. The many subsectors involved makes the construction industry unique (Ashworth and Perera, 2015; Mokhtariani et al., 2017). The subsectors include design (consultancy inclusive), construction, installation, and construction management. The subsectors' activities include survey and design, construction, installations, construction management/supervision and consultancy of construction projects. The global construction market is projected to grow by 85% to US$15.5 trillion by 2030 (Regona et al., 2022). For this to happen, technological advancements will be pertinent. Some relevant characteristics are as follows.

- Manufacturing differs from the construction industry because of the systematic nature and complexity of construction, meaning that 'systems thinking' is needed in construction activities.
- The industry is measured by on-site production.
- The products from the industry contribute to industrialisation.
- In the 21st century, products are shifting from labour-intensive to technology-based via advanced construction digitalisation.
- The industry is a conglomerate of industries or a meta-industry.
- Uncertainties are high in construction projects because of the turbulent environment, requiring semi-predictable configurations of expertise and supply industries.
- No two construction projects are alike.
- Construction activities are conducted on the site where end users will use the product.
- Construction activities are influenced by weather and many tasks cannot be conducted in controlled factory conditions.
- The end products are immobile.
- The 'production' procedure includes a multi-faceted combination of trades, skills and materials.
- The sector is more of a service than a manufacturing industry because it renders service activities.
- The sector includes a few very large construction companies and larger numbers of small firms.
- Many of the companies that produce construction products are small or medium sized businesses.

- The industry is embracing innovation, including digitalisation and fast growth century (industrialisation of construction).
- The industry is embracing and growing an inter-related ecosystem of software and hardware attributes (sustainable construction).

1.3. Major stakeholders in the building industry

Many construction projects involve a variety of stakeholders. Stakeholders in this context imply individuals or any group who can influence the achievement of a corporate goal. Besides influencing to achieve a goal, a stakeholder may have a vested interest or share in an undertaking. In a building project, numerous stakeholders are involved, including organisations, the natural environment, government agencies and ministries, guests to the facility, community representatives, insurance firms, financial institutions, construction workers, legal authorities, service providers, suppliers, subcontractors, shareholders, designers, project managers, users of facilities, managers and owners/clients (Jin *et al.*, 2017). This section focuses on the major stakeholders in a building project – the clients/employers, design teams and contractors/housing developers. Their actions or decisions can influence the progress of a construction project. And these actions or decisions are in turn influenced by the stakeholders' main responsibilities, attributes, experience, approach to risk management and professional background. Clearly, this can sometimes be complicated. The construction sector can be subgrouped into three major areas (Ebekozien and Aigbavboa, 2024) – building, civil engineering and heavy engineering work. Each group complements the others and there are relationships between them. Figure 1.1 shows the major stakeholders in the building industry, which are the design team, client employer and housing developer/contractor.

1.3.1 Design team

The design phase of a building project is fragmented into several specialties and requires information flow to achieve the goal (Herrera *et al.*, 2020). If not mitigated, poor interactions within the design team can lead to poor building project performance. This implies that the performance of a building project can improve when the design team's performance is improved (Ashworth and Perera, 2015; Senaratne and Gunawardane, 2015). Thus, the design team's effectiveness is key in integrated building project delivery. Successful design team management is pertinent via collaboration and good teamwork practices – from the pre- to post-construction phases – for timeliness, cost-effectiveness and quality of building projects.

1.3.1.1 Quantity surveyor

A quantity surveyor (QS) is a construction cost manager who offers financial services and economic consultations for projects. Background training in construction technology, information management, construction economics and contractual law gives a QS an important place in the industry. Quantity surveying involves project construction cost management, procurement and contractual matters. Understanding the dynamics of construction costs is critical to the QS. The Nigerian Institute of Quantity Surveyors (NIQS, 1998) defines a QS as

> a cost and procurement management expert concerned with financial probity and achieving value for money in the conceptualisation, planning, and execution of building, civil, and other heavy engineering projects

Figure 1.1 Major stakeholders in the building industry

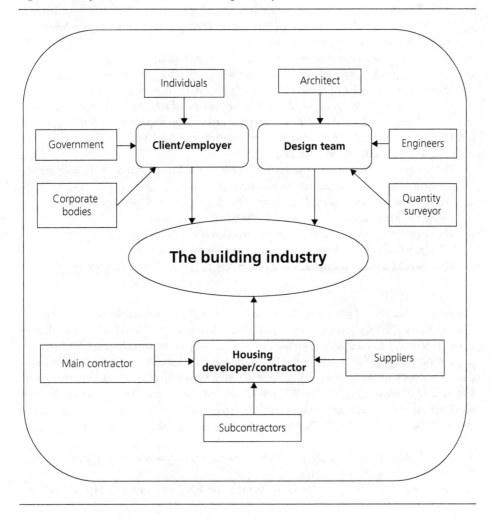

A QS is the cost manager and engineer in construction projects. Thus

> quantity surveying is primarily concerned with the detailed calculations and assessment of quantities of materials and labour required for all construction activities such as building, civil and heavy engineering (NIQS, 1998)

The QS should be acquainted with construction projects and allied subjects. As a cost adviser or technical accountant on site, a QS is involved in a building project from commencement to its ultimate financial settlement. One function of a QS is to prepare a bill of quantities (BOQs). Other contract documents prepared are the specification of materials and workmanship or preambles, preliminaries bills and contract conditions (Ashworth *et al.*, 2013; Chandramohan *et al.*, 2022). The duties of a QS can be summarised as follows.

- The QS is available to guide the employer and architect on the proposed construction's cost-effectiveness, including a sustainable advisory role.
- The QS advises on the financial aspect of construction projects. This includes heavy engineering, civil and building projects.
- The QS prepares an approximate construction cost estimate.
- The QS prepares BOQs. Where this is impossible due to the absence of adequate information, schedules of prices or bills of approximate quantities are prepared.
- The QS prices the BOQs or schedules of prices in line with best global practices.
- The QS prepares the documents necessary for procuring competitive tenders.
- The QS advises on the form of contract and payment terms. This is important to mitigate or prevent construction disputes.
- The QS evaluates the construction work in progress and recommends to the architect the payment to be issued on the interim certificate.
- The QS conducts the technical checking of accounts presented by the construction contractor. These accounts are based on the construction contractor's costs.
- The QS provides expert evidence on construction disputes.
- The QS prepares and settles construction contractors' final accounts.
- The QS is an expert in construction procurement because of their expertise and skills.

1.3.1.2 Architect

For decades, the role of the architect (building designer) has continued to evolve and shift (Burr and Jones, 2010). The 21st century has opened up new opportunities for technological advancement and collaboration with the design team to achieve specific goals. The International Union of Architects (IUA, 2011) describe architecture as involving '…everything that influences how the built environment is planned, designed, made, used, furnished, landscaped, and maintained…' and so '…Architectural education constitutes some of the most significant environmental and professional challenges of the contemporary world…' (Feria and Amado, 2019; IUA, 2011). The duties of an architect can be summarised as follows.

- The architect should have discussions with the client, collect briefs and visit the site to facilitate the preliminary design.
- The architect should crystallise the preliminary design (brief) into a factual foundation for the building project.
- The architect should ensure that the preliminary design complies with statutory requirements.
- After checking with other consultants, a sketch design is produced and the client's approval is sought.
- The architect checks with the QS to see if the cost matches the budget.
- In the detailed design phase, the architect considers the client's requests in consultation with other team members.
- The architect heads the design team – this is not a right, but is determined by many factors (e.g. the scope of work and the client's discretion).
- The architect prepares tender/contract drawings and coordinates with other consultants.
- The architect is fully responsible for ensuring what has been designed is actually constructed, if included in the agreement with the client.
- The architect issues certification of work done and interim certificates at agreed intervals.

- The architect issues a practical completion certificate to the contractor/housing developer at practical completion.
- The architect also issues a certificate of making good and where defects/faults, if any, have been made good. This is usually after a defect liability period of 6 months.

1.3.1.3 Engineers

Electrical, mechanical and structural engineers are engaged in most building projects during the design phase. Engineers play a crucial role in achieving a successful building project (Gaur, 2023), including planning (during design), monitoring and controlling (during construction) the building project. Collaboration and information sharing are key among the engineers and other stakeholders to positively impact project costs and time performance. Briefly, the duties include the following.

- The engineer works with other team members, including the architect and the quantity surveyor, to achieve the project goal.
- The engineer ensures that the proposal is adequate and can be realised.
- The engineer is responsible for generating relevant detailed calculations, schedules and specifications to enhance other team members in advisory decisions to the client.
- The engineer ensures that the scheme satisfies local and international statutory requirements.
- The engineer supervises the project and modifies the scheme when necessary. This could be optional if not covered in the contract with the client.
- For some construction projects, especially in civil engineering, the civil engineer becomes the main consultant and the services of an architect may not be required.

1.3.2 Client (employer)

The client could be an individual, an organisation or a government agency (federal, state or local government). Ideas for projects usually emanate from a client, except in speculative development. Thus, the client plays an important role in developing the right conditions for project performance via innovation (Lindblad and Guerrero, 2020). According to Lindblad and Guerrero (2020), clients often work in partnership to address the issue of uncertainty and other emerging matters in construction. This is in agreement with Loosemore (2015), who noted that, in some instances, housing developers/contractors depend on clients with regard to innovation since project constellations are involved. Briefly, the client's main duties include the following.

- The client's responsibility starts with conceptualising a design brief. The brief sets out the components of the project, including the aim of the project.
- The client needs to employ the services of experienced consultants to handle projects. Sometimes, this action is taken too late and the project suffers the consequences.
- To comply with the contract terms, the client should have a feasible budget and predictable cash flow. Many clients have defaulted, including public projects in Nigeria, Ghana and South Africa. Examples of failed projects due to funding issues in Nigeria included the Ilorin–Jebba–Mokwa–Birni Gwari expressway, the Enugu–Onitsha expressway and the Benin–Auchi expressway (Olawoyin and Ukpong, 2018). In 2017, it was reported by Ghana Web that most suspended or abandoned construction projects

are due to non-payment of work certificates, costing the state about GhC,μ 30 million (£1.8 million) (Kuoribo et al., 2021).
- The client should settle all payment certificates as they are due.

1.3.3 Contractor

Contractors are project-based firms. They are decentralised where construction projects are managed (Haglund and Rudberg, 2023). Contractors range in size from one-person outfits to large multi-national conglomerates. In developing countries, selection and recruitment processes are fraught with issues (Evarist et al., 2023). If not checked, unqualified construction workers can result in poor project delivery. Contracting firms not only vary in size, but also in capability and reliability. Some contractors have teams that can handle all facets of construction, while others rely on the services of subcontractors for specialist jobs. Generally, the following are some of the expected duties from an engaged contractor.

- On 'winning' a contract, the contractor should accept to handle it.
- The contractor should conduct a site investigation for abnormal conditions before submitting a final commercial bid.
- The contractor should give the client and consultants duplicates of the priced BOQs, if required, for ease of cost control.
- The contractor should raise queries on drawings in cases of discrepancies during the pre-contract and contract phases.
- The contractor should host regular site meetings and update the participants.
- The contractor should provide site facilities for consultants, other site staff and support subcontractors.
- The contractor should provide consultants with enough information to help them monitor progress.
- The contractor should take out insurance cover as required by the contract agreement.
- The contractor should adhere to the contract conditions, rectify defects and complete the work on schedule.
- Nominated subcontractors and suppliers have obligations to the contractor.

1.4. The construction industry as an economic regulator

Since the Second World War, many developed and developing countries' governments have used the construction industry as an economic regulator. Governments are major clients and links to other industries, and the high proportion of public expenditure in total construction output is another factor to consider (Ashworth and Perera, 2015; Ball, 2014). Expenditure on construction-related products within limits thus becomes a means of reflation, making it a soft option for the treasury knife (i.e. public expenditure cuts). Governments remain major clients in the construction sector in many developing countries, including Nigeria and Ghana. Some scholars have argued that the construction sector is an economic regulator but this should not be exaggerated – governments may concede or drop projects for different reasons. For example, decreasing public sector borrowing often makes an impact. Public expenditure cuts can occasionally have a large effect on construction, yet are frequently accompanied by different measures (Abdullahi and Bala, 2018; Olanipekun and Saka, 2019). In addition, the cancellation of awarded contracts after extensive pre-planning and design can be dangerous to the system and whether these can be referred to as an example of regulation is debatable.

The issues within and surrounding the construction sector are debatable, but outside the scope of this book.

The health of the construction sector ought to be high on the political agenda, given the employment opportunities and its contribution to the economy. However, the situation is different in many developing countries. Ofori (2012) pointed out that a country can only benefit from the economic stimulus that the construction sector can offer if the industry is capable. These benefits are not feasible in Nigeria and Ghana because of lax government attitudes towards the industry. The ability to increase activity in other industries of domestic economies is vital to enhance the benefits of the construction sector to the economy. Over the years, governments have mediated construction markets by, for example,

- creating stimuli through funds and subsidies
- providing construction grants for projects in locations of high unemployment
- offering construction incentives for specific projects (e.g. low-cost housing developments)
- granting tax breaks for construction companies on specific projects
- creating prospects for construction by permitting the extensive scope of activities in confined regions
- having direct involvement in projects' development, repair or maintenance through direct labour.

1.5. Effects of government action on the construction industry

Government actions influence the construction sector outputs, globally. Studies have shown that a fluctuating workload hurts the sector concerning the management and planning of resources because major construction jobs come from governments. Olanipekun and Saka (2019) note that government policies affect the construction industry, with policies either restraining or stimulating the economy but with a few exemptions. When government regulations restrain operations, the volume of public construction projects is reduced and job losses are increased. The adverse effects of such policies are

- unemployment of construction workers
- smaller firms being forced out of the industry
- reluctance of large firms to invest in machinery with evolving innovations and technologies
- reluctant cooperation of subcontractors and suppliers of construction materials
- increasing difficulty in engagement with workers
- increased costs and reduced efficiency because of the absence of work continuity.

The following are repercussions of government policy.

- *Government as client*. As in other countries, the Nigerian government is the major client for basic infrastructure projects (e.g. electrification, roads, buildings, expanding towns, etc.). Thus, if government spending on roads, health care centres and schools and so on were to be cut, this would negatively affect the construction sector. The priorities of

governments can also change. For example, moving funds for construction projects to address other needs will disadvantage the construction industry.
- *Control.* Governments can influence the locations of industries or new construction projects in certain ways. This influence is more pronounced when the government is directly or indirectly involved in the project by providing construction subsidies, grants, tax breaks or other incentives.
- *Monetary policy.* Governments employ different methods to adjust interest rates, such as open market operations, bank rates and higher purchase limitations. The outcomes of these can discourage some contractors/housing developers and clients from investing in the sector.
- *Taxation.* Taxation is a key instrument in government fiscal policies because it is charged on income and capital. An increase in capital tax on the value of a building/project or profits made from business transactions of the building can thus lead to diminished interest. Likewise, increased property income might increase interest in the new project.
- *Fiscal policy.* Governments can encourage the degree of monetary events by controlling expenditure. This affects the construction industry because governments are the major clients. However, in times of rising unemployment, increasing spending on public works will create jobs.

1.6. The history of building economics

The history of building economics is long, but became a distinct ground in the 1970s. A few decades before the mid-1970s, basic infrastructure such as roads, hospitals, schools, residential buildings, offices and markets were needed to meet population expansion and urbanisation. Meeting these goals with scarce resources required extra engagement that would yield value for money. This scenario birthed the application of basic economics principles in construction. Economics is about the choice from possible options, bearing in mind that resources can be scarce and value productivity is pertinent. Any decision for a selection should be justified. Thus, building or construction economics is a branch of general economics with a focus of using economic theories to achieve the needs of the construction industry (Ashworth and Perera, 2015). In economies, construction signifies capital stock and expenditure (Bon, 2001). In Nigeria, quantity surveying programmes in higher education institutions in the 1970s enhanced research on the subject of building economics. However, the history of construction economics can be traced back to quantity surveying practices in the UK as a component to assist decision-makers with building economic issues on a daily basis. In addition, the economic forecasting aspect of building economics cannot be over-emphasised. It remains a preoccupation of the building process and includes life-cycle costing. This came to the limelight in the mid-20th century with contractors using approximate estimating/forecasting to present budgets to employers.

Building economics and quantity surveying practices are interconnected. Building economics principles show construction practitioners how to apply economics concepts to decisions concerning the design, location, construction, operation, management, rehabilitation and demolition of buildings. Studies have revealed that use of these methods/tools can lower construction costs and boost profits. This is a germane point – clients/employers wish to increase profits and lower construction costs without altering the quality. Achieving this goal requires an understanding of building economics concepts/tools. Some of the tools/concepts will be discussed in subsequent chapters of this book. The subject of building economics is increasing in academic curricula and among professional practitioners. This is because of the economic process it contains and how it offers a systematic framework to enable quick economic decisions to be made. Forecasting, budgeting, controlling, accounting and evaluating will always be associated with construction costs in order to achieve value for money, but the future for building economics looks like the following.

- The development of better or more economic tools may remain high on the agenda.
- Digitalisation will be embraced for the capture and analysis data.
- Sustainable construction will advance in order to boost contractors/housing developers' profits and reduce maintenance costs.
- Economic forecasts should shift to decisions regarding the use of capital rather than investment decisions.
- To develop building economics further, more solid theoretical underpinnings of construction economics are needed – from conception to demolition.
- Corporate real estate and facilities management should be embraced into the subject.

1.7. Construction economics and its scope

Construction or building economics is a branch of general economics. There is no universal definition of what construction economics is. It is concerned with applying economic tools to maximise profits and reduce project construction costs. In general, economics can be described as the determination of how scarce resources are and how they should be allocated between all possible uses. The scope of economics implies identifying the subject matter of economics and deciding whether economics is a science. Thus, the scope of economics is any area that economics touches as a discipline. Various definitions of economics have indicated this, but there seem to be diverse views on the definition of the subject. According to Myers (2022), the study of economics centres on scarcity and how individuals or society try to rationalise scarce resources in the best ways for maximum satisfaction. Therefore, the basic concepts in economics are scarcity, choice and cost of opportunity. From a simple definition of economics, construction economics is about people's needs for housing and the right conditions in which to live.

Construction or building economics embraces the following.

- Construction cost minimisation to achieve the requirements of the client/employer. This implies ensuring that the final account resembles the estimate within the specific completion time and a reasonable cost is maintained (Myers, 2022). Understanding and meeting the client's wants and needs, including project design and maintenance at reasonable cost, is germane in construction economics (Ashworth and Perera, 2015).
- Besides forecasting cash flow, predicting the total construction cost, investment analysis and evaluating construction site productivity cannot be over-emphasised. This includes the maintenance and running costs of buildings and their components.
- Construction economics also involves macroeconomic activities. This includes studying how the industry relates to the rest of the economy regarding applications of sectoral input/output, the impact of money supply on construction output and the impact of construction output on GDP. The relationship between shape and space

cannot be over-emphasised because of its impact on overall cost designs (Ashworth and Perera, 2015).
- The initial cost assessment of a construction project and the environment surrounding the project's location are pertinent in decision-making. The outcomes of these may influence the planning and the general amenities of the proposed construction project.
- Methods for controlling costs will vary depending on the project type and the nature of the client. Understanding the various control methods will enhance the growth of construction economics. This is a component of the scope, with emphasis on initial building estimates and cost controlling.
- Estimating the life of construction projects and materials, especially sustainable construction materials, is key. This includes the role of digitalisation in construction activities in the 21st century, especially in developing countries such as Nigeria, Ghana and South Africa.

1.8. Summary

This chapter introduced readers to the structure of the construction industry and its inter-relationship with industry stakeholders. The importance of the industry as an economic regulator and its impact on national economies were then described. An overview of the key functions of main stakeholders in the industry was given. The chapter also highlighted the effects of government action on the industry, the history of building economics and its scope for the built environment. The next chapter provides an in-depth description of approximate estimating, the various forms of approximate estimating and design variables. The chapter concludes with economic comparisons of alternative designs.

REFERENCES

Abdullahi M and Bala K (2018) Analysis of the causality links between the growth of the construction industry and the growth of the Nigerian economy. *Journal of Construction in Developing Countries* **23(1)**: 103–113.

Akdag SG and Maqsood U (2019) A roadmap for BIM adoption and implementation in developing countries: the Pakistan case. *Archnet-IJAR: International Journal of Architectural Research* **14(1)**: 112–132, https://doi.org/10.1108/ARCH-04-2019-0081.

Asare E, Owusu-Manu DG, Ayarkwa J, Edwards DJ and Martek I (2022) Thematic literature review of working capital management in the construction industry: trends and research opportunities. *Construction Innovation* **23(4)**: 775–791, https://doi.org/10.1108/CI-09-2021-0177.

Ashworth A and Perera S (2015) *Cost Studies of Buildings*. Routledge, London, UK.

Ashworth A, Hogg K and Higgs C (2013) *Willis's Practice and Procedure for the Quantity Surveyor*. Wiley, London, UK.

Ball M (2014) *Rebuilding Construction (Routledge Revivals): Economic Change in the British Construction Industry*. Routledge, London, UK.

Boadu EF, Wang CC and Sunindijo RY (2020) Characteristics of the construction industry in developing countries and its implications for health and safety: an exploratory study in Ghana. *International Journal of Environmental Research and Public Health* **17(11)**: 4110.

Bon R (2001) The future of building economics: a note. *Construction Management & Economics* **19(3)**: 255–258, https://doi.org/10.1080/01446190010020354.

Burr LK and Jones BC (2010) The role of the architect: changes of the past, practices of the present, and indications of the future. *International Journal of Construction Education and Research* **6(2)**: 122–138, https://doi.org/10.1080/15578771.2010.482878.

Chandramohan A, Perera SKAB and Dewagoda GK (2022) Diversification of professional quantity surveyors' roles in the construction industry: the skills and competencies required. *International Journal of Construction Management* **22(7)**: 1374–1381, https://doi.org/10.1080/15623599.2020.1720058.

Delgado DMJ, Oyedele L, Ajayi A *et al.* (2019) Robotics and automated systems in construction: Understanding industry-specific challenges for adoption. *Journal of Building Engineering* **26**: 100868, https://doi.org/10.1016/j.jobe.2019.100868.

Feria M and Amado M (2019) Architectural design: sustainability in the decision-making process. *Buildings* **9(5)**: 135–147.

Ebekozien A (2020) Corrupt acts in the Nigerian construction industry: is the ruling party fighting corruption? *Journal of Contemporary African Studies* **38(3)**: 348–365. https://doi.org/10.1080/02589001.2020.1758304.

Ebekozien A (2022) Construction companies' compliance to personal protective equipment on junior staff in Nigeria: issues and solutions. *International Journal of Building Pathology and Adaptation* **40(4)**: 481–498.

Ebekozien A and Aigbavboa C (2024) *Professional Practice for Quantity Surveyors in the 21st Century*. Taylor & Francis, London, UK.

Evarist C, Luvara MGV and Chileshe N (2023) Perception on constraining factors impacting recruitment and selection practices of building contractors in Dar Es Salaam, Tanzania. *International Journal of Construction Management* **23(12)**: 2012–2023, https://doi.org/10.1080/15623599.2022.2031556.

Gaur S (2023) Importance, roles and responsibilities of a planning engineer in the Indian construction industry. *European Journal of Theoretical and Applied Sciences* **1(3)**: 328–340.

Haglund P and Rudberg M (2023) A longitudinal study on logistics strategy: the case of a building contractor. *The International Journal of Logistics Management* **34(7)**: 1–23, https://doi.org/10.1108/IJLM-02-2022-0060.

Herrera RF, Mourgues C, Alarcon LF and Pellicer E (2020) Understanding interactions between design team members of construction projects using social network analysis. *Journal of Construction Engineering and Management* **146(6)**: 1–13, https://doi.org/10.1061/(ASCE)CO.1943-7862.0001841.

Jin X, Zhang G, Liu J, Feng Y and Zuo J (2017) Major participants in the construction industry and their approaches to risks: a theoretical framework. *Procedia Engineering* **182**: 314–320.

Kuoribo E, Amoah P, Seidu S and Abdulai SF (2021) Highlighting the effects of uncompleted/abandoned construction projects in Ghana. *Proceedings of The IDoBE International Conference, London, UK.*

Lindblad H and Guerrero JR (2020) Client's role in promoting BIM implementation and innovation in construction. *Construction Management and Economics* **38(5)**: 468–482.

Loosemore M (2015) Construction innovation: fifth generation perspective. *Journal of Management in Engineering* **31(6)**: 1–9.

Mokhtariani M, Sebt MH and Davoudpour H (2017) Characteristics of the construction industry from the marketing viewpoint: challenges and solutions. *Civil Engineering Journal* **3(9)**: 701–714.

Myers D (2022) *Construction Economics: A New Approach*. Routledge, London, UK.

NIQS (Nigerian Institute of Quantity Surveyors) (1998) *Directory of Members and Quantity Surveying Firms*, 4th edn. NIQS, Lagos, Nigeria.

Nwachukwu CC and Nzotta MS (2010) Quality factors indexes: a measure of project success constraints in a developing economy. *Interdisciplinary Journal of Contemporary Research in Business* **2(2)**: 505–512.

Ofori G (2012) *Developing the Construction Industry in Ghana: The Case for a Central Agency*. National University of Singapore, Singapore.

Ofori G (2015) Nature of the construction industry, its needs and its development: a review of four decades of research. *Journal of Construction in Developing Countries* **20(2)**: 115–135.

Ofori-Kuragu JK (2013) *Enabling World-class Performance in Ghanaian Contractors: A Framework for Benchmarking*. PhD thesis, Kwame Nkrumah University of Science and Technology, Kumasi, Ghana.

Olanipekun AO and Saka N (2019) Response of the Nigerian construction sector to economic shocks. *Construction Economics and Building* **19(2)**: 160-180.

Olanrewaju OI, Chileshe N, Babarinde SA and Sandanayake M (2020) Investigating the barriers to building information modeling (BIM) implementation within the Nigerian construction industry. *Engineering, Construction and Architectural Management* **27(10)**: 2931–2958, https://doi.org/10.1108/ECAM-01-2020-0042.

Olawoyin O and Ukpong C (2018) Special Report: Amidst Inadequate Funding, Buhari's Road Construction Projects Across Nigeria Continue. See https://www.premiumtimesng.com/news/headlines/289368-special-report-amidst- inadequate-funding-buharis-road-construction-projects-across-nigeria-continue- 1.html?tztc=1 (accessed 21/03/2024).

Regona M, Yigitcanlar T, Xia B and Li MYR (2022) Opportunities and adoption challenges of AI in the construction industry: A PRISMA review. *Journal of Open Innovation: Technology, Market, and Complexity* **8(1)**: 45, https://doi.org/10.3390/joitmc8010045.

Senaratne S and Gunawardane S (2015) Application of team role theory to construction design teams. *Architectural Engineering and Design Management* **11(1)**: 1–20, https://doi.org/10.1080/17452007.2013.802980.

IUA (International Union of Architects) (2011) UNESCO/UIA Charter for Architectural Education. See https://etsab.upc.edu/ca/shared/a-escola/a3-qualitat/validacio/0_chart.pdf (accessed 21/03/2022).

Unegbu OCH, Yawas DS and Dan-asabe B (2022) An investigation of the relationship between project performance measures and project management practices of construction projects for the construction industry in Nigeria. *Journal of King Saud University – Engineering Sciences* **34(4)**: 240–249, https://doi.org/10.1016/j.jksues.2020.10.001.

WEF (World Economic Forum) (2018) Shaping the Future of Construction – Future Scenarios and Implications for the Industry. WEF, Geneva, Switzerland. See http://www3.weforum.org/docs/Future_Scenarios_Implications_Industry_report_2018.pdf (accessed 20 March 2022).

Winch GM (2003) How innovative is construction? Comparing aggregated data on construction innovation and other sectors – a case of apples and pears. *Construction Management and Economics* **21(6)**: 651–654, https://doi.org/10.1080/0144619032000113708.

World Bank (2013) *The World Bank Annual Report 2013*. The World Bank, Washington, DC, USA.

Principles of Basic Construction Economics in the 21st Century

Andrew Ebekozien and Clinton Aigbavboa
ISBN 978-1-83549-841-5
https://doi.org/10.1108/978-1-83549-838-520241002
Emerald Publishing Limited: All rights reserved

Chapter 2
Approximate estimating and design variables in the construction industry

2.1. Approximate estimating

Approximate estimating, sometimes referred to as probable costing, is a challenging task in building economics. It is an attempt to forecast the cost of a building project very early in the design stage, often before detailed designs are available. It is conducted to find out an approximate construction cost of a proposed project (Seeley, 2010), with the main purpose of generating a forecast or preparing a probable cost of a proposed construction project before the detailed design and other contract documents are prepared. A professional quantity surveyor (QS) needs to have vast experience to develop a reliable probable construction cost. This task is not achieved in isolation without the cooperation of other project members, including the client, architect and engineers. Several methods can be used, the choice of which could be influenced by (*a*) the time allocated and the information available, (*b*) the QS's expertise and experience in approximate estimating and (*c*) the availability of cost data and other related data.

Besides visiting the site for physical observations to aid a preliminary estimate, an experienced QS would insist on a detailed site survey report to guide key assumptions such as satisfactory load bearing capacity, adequate groundwater level, fill-free site, levelled locations and so on. It is in the client's best interest to have a probable cost that is as accurate as possible. This will help the client to make a decision on whether to continue with the project or not. In addition, clients will have different requirements. For example, the client

(a) wishes to have a certain building built with cost limits
(b) wishes to have a building of a certain type, size and specification but with no initial fixed budget (Ashworth and Perera, 2015) (e.g. an individually designed and built private house)
(c) wants the maximum floor area for a specific sum of money (e.g. a lecture hall).

In (*a*), both the area (size) and cost are fixed. In (*b*), the area (size) is fixed but the cost is not initially fixed. In (*c*), the area (size) has yet to be determined but the cost is fixed. Each of the above scenarios poses different problems for the architect and the QS.

2.2. Forms of approximate estimating

2.2.1 Unit method

This is one of the simplest methods of approximate estimating. It involves allocating a cost to each accommodation unit of the structure or a cost per place. There is a correlation between the cost of a building and the units it accommodates. In this context, the units are variables

15

that the proposed building will use. Although the approach looks easy, it requires high expertise and skills to select suitable rates. The simplicity of the method is an advantage, but care should be taken regarding its precision. The QS needs to consider some variables before arriving at the final unit rate, which include the market condition, specification changes that are cost-related and varying site conditions. These factors could influence the unit rate. Examples of building project unit rates are shown in Table 2.1. The advantages and disadvantages of the method are shown in Table 2.2.

2.2.2 Cube method

This is the traditional method of approximate estimating. It is hardly used by practitioners in the 21st century, but was widely employed between the First and Second World Wars. The advantages and disadvantages of the method are shown in Table 2.3.

The Royal Institute of British Architect (RIBA) gives the following guidelines, as cited in Ashworth and Perera (2015).

- The length, width and height of the building dimensions (external) are multiplied and expressed in cubic metres (m^3).

Table 2.1 Examples of building project unit rate

Building project	Unit rate
Secondary school	Cost/student
Primary school	Cost/pupil
Stadium	Cost/seat
Student hostel	Cost/bed or cost/student
Hospital	Cost/bed
Cinema	Cost/seat
Church	Cost/seat or cost/member
Mosque	Cost/seat or cost/member

Table 2.2 Advantages and disadvantages of the unit method

Advantages	Disadvantages
<ul><li>It is simple to use (compared with the cube method</li><li>It is quick to apply. For example, for a proposed stadium, the previous cost/seat rate used for a similar stadium can be applied</li></ul>	<ul><li>Lacks precision because building height and other factors (e.g. foundation type) are not considered</li><li>Unit rates are expressed within a range of prices</li><li>Involves engaging a more reliable estimating method to arrive at acceptable unit rate</li></ul>

Table 2.3 Advantages and disadvantages of the cube method

Advantages	Disadvantages
• Useful for estimating service-related activities such as air conditioning and heating • It is a simple approach to determine building volume	• Cost relates to building shape • The client is not preview to the amount of functional floor area • The approach is not satisfactory for universal applications • It has deceptive simplicity • It fails to make provisions for cost-related variables such as the number of storeys, storey heights and plan shape

■ Regarding the building height, the approach is determined by the roof type. For a pitched roof, it is taken from the top of concrete foundation to halfway up the roof. For a flat roof, it is taken as 600 mm above the roof. For an occupied roof space, it is taken up to 0.75 of the way up the roof. If there is a parapet in the flat roof, it is taken up to 600 mm above the parapet. These concepts are illustrated in Figure 2.1.

For multi-storeyed buildings, the QS must consider more assumptions, such as (*a*) per 1 m³, the floors are equal in cost and (*b*) the basement cost is 60% of the cost of different floors.

Using the cube method, for a storey height of 4 m × 0.80 foundation + 0.60 m above the flat roof, the estimate is:

$4 + 0.8 + 0.6 = 5.40$ m (height)
$\rightarrow 744 \times 5.4 = 4017.6$ m³
Cubic rate = ₦75/m³
Total approximate estimate = 4017.6×75 = ₦301 320

Figure 2.1 Floor plan and section of pitch and flat roofs

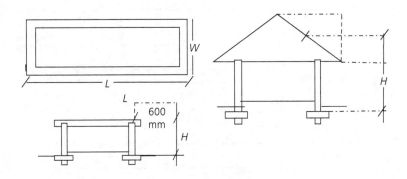

2.2.3 Superficial or floor area method

This is a common approach used by practitioners in calculating an approximate estimate for a proposed project. Calculating the estimate using this method is simple and understandable to the average investor. The floor area is measured and multiplied by the cost/m². The measurement is taken between the internal faces of the stairs, columns, partitions, internal walls and the like. However, if the employer/client expresses requirements concerning functional area, non-usable space and circulation would be added to the area. The cost/m² used is generated from cost analyses of past completed building projects of similar construction technique, storey height and plan shape. With this method, the available rates need to be subjectively adjusted to take account of prevailing market conditions.

The following is an extract of the approximate estimate of a lecture hall using the superficial area method.

Total area = 744 m²
Lecture hall cost/m² = ₦405 (US$1 = ₦1410)
Total approximate estimate = 744 m² × ₦405 = ₦301 320

The advantages and disadvantages of the method are shown in Table 2.4.

In the following example, consider a building 12 m × 25 m in plan, with a wall thickness of 225 m (see Figure 2.2). The rate/m² used in the example (₦350/m²) would be obtained from analysis of one or more buildings similar in type and quality of finish.

Table 2.4 Advantages and disadvantages of the floor area method

Advantages	Disadvantages
• The calculation is easy. • An average building client understands how cost is expressed. • There are several sources to obtain rates. • There is a correlation between the floor area and the building items.	• Cost relates better with building shape than floor area. This is because building shapes take account of the total height and the number of storeys among which its total area is allocated. • Adjusting cost-related variables is not friendly. • This approach is not satisfactory for universal applications.

Figure 2.2 Sketch plan of proposed building

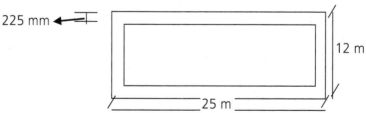

The internal floor dimensions are

> 12 m − (2 × 225 mm) = 11.55 m
> 25 m − (2 × 225 mm) = 24.55 m
> Floor area = 24.55 × 11.55 = 283.55 m^2
> Assuming two floors of the same area
> Total floor area = 283.55 × 2 = 567.10 m^2
> Estimated cost of building = 567 m^2 × ₦350/m^2 = ₦198,450 (US$1 = ₦1410)

2.2.4   Perimeter method

The uniqueness of this method is that there is a deviation from the previous three approximate estimating methods because the method combines the building's perimeter and the floor area. In the economic design of a building, the floor area is pertinent. This method is better than the other methods described so far because of the inclusiveness of the building's perimeter. However, the scarcity of cost databases have created a resistance from QSs to explore this method. The advantages and disadvantages of the method are shown in Table 2.5.

2.2.5   Storey-enclosure method

To address the drawbacks of the first four approximate estimating approaches, the storey-enclosure approach emerged and provides better accuracy and precision. It evolved to address the issues associated with the single-price approaches. Seeley (2010) identified key variables that should be considered in calculating an approximate estimate using this method. These include the building plan shape, the total floor area, the vertical position of floors, the storey heights, the overall building height and the extra costs of offering usable floor below ground. The advantages and disadvantages of the method are shown in Table 2.6.

Table 2.5 Advantages and disadvantages of the perimeter method

Advantages	Disadvantage
• Better than the superficial methods because of the inclusiveness of perimeter • Provides better precision for cost planning	• No cost databases with appropriate rates

Table 2.6 Advantages and disadvantages of the storey-enclosure method

Advantages	Disadvantages
• This approach performs better than the other methods in terms of accuracy and precision • Cost-influencing variables such as storey height, plan shape and so on are considered	• The weightings used are subjective • The attributes do not relate to clients' accommodation requirements

2.2.6 Approximate quantities

Most QSs use this approach because the pattern of preparation is similar to that for the bill of quantities (BOQ) approach. The presentation is given in terms of composite items, so this method offers a more detailed approximate estimate. The items are measured or grouped using the typical billed mechanism with emphasis on items that are of cost relevance. Besides providing detailed information, the method is reliable. However, it is time-consuming. There are no measurement rules applicable to a formal BOQ preparation. However, approximate quantities are different from a bill of approximate quantities.

For example, using approximate quantities, a building roof may be measured as follows. Three-layer bituminous felt on 50 mm pre-felted woodwool floor, with firings, on 75×150 mm hardwood joists at 300 mm centres, with a vapour barrier and 100 mm fibreglass insulation. The all-in rate for this above is as follows.

Total area is $100 \, m^2$
Three-layer felt = ₦200
50 mm woodwool = ₦300
50×75 mm firings = 4m@₦150 = ₦600
75 mm $\times$ 150 mm joists = 4m@₦100 = ₦400
Vapour barrier = ₦50
Fibreglass = ₦150
Total = ₦1700
Plus sundries (5%) = ₦85
Total cost = ₦1785/m^2 (US$1 = ₦1410)

The advantages and disadvantages of the method are shown in Table 2.7.

2.2.7 Elemental estimating

In this method, the first stages of cost planning are used to generate approximate project costs. Costs of similar project are analysed. The costs can be expressed in three ways for each element: the gross index floor area, the cost/m^2 and the total cost. The cost/m^2 is more convenient for practitioners. To calculate the cost/m^2, the total building cost is divided by the total floor area. This sum is assigned to each element using previous cost analyses as a guide.

2.2.8 Analytical estimating

A contractor's estimator may adopt this method to determine individual rates for measured items on the BOQ. Each item is analysed – materials, labour and plant. This method also

Table 2.7 Advantages and disadvantages of the approximate quantities method

Advantages	Disadvantages
• Provides a more reliable approximate estimate	• Takes more time
• Offers greater precision	• Requires more details

considers the variables that could influence the construction cost, including site location and the size, shape and height of the building. The analytical method is applicable in situations where a new construction technique is envisioned. What this implies is that it is mostly applicable where there are no available or existing data, so the estimator does not have any other choice but to use this method.

2.2.9   Cost models
This is one of the new generation of methods of forecasting construction costs. It is a statistical method and can be used with spreadsheets or other computer aids. The easier the method of measuring cost models, the more difficult, and hence the more inaccurate, will be the pricing. Unless rates are easily obtainable and currently available, the method will find little acceptance in practice. However, in this method, a model or formula is generated that defines data collation regarding costs or price. For example

$$y = x_1 + x_2 + x_3 + x_4 + x_5 + x_6 + x_7 + x_n$$

where y is the cost of the building, x_1 is the cost of the substructure, x_2 is the cost of the walls, x_3 is the cost of the roof, x_4 is the cost of windows and doors, x_5 is the cost of finishes, x_6 is the cost of mechanical, electrical and plumbing installation and x_7 is the cost of external works.

2.2.10   Financial methods
In this method, a cost limit is assigned to the design. The design team members, especially the architect and the QS, have to ensure that the design can be constructed within the assigned cost limit.

2.3.   Design variables
There are plenty of texts that discuss the effects of various design variables on construction costs (e.g. Ashworth *et al.*, 2013; Ball, 2014; Genc, 2023; Lowe *et al.*, 2006, 2007; Orzeł and Wolniak, 2022; Sarker *et al.*, 2012; Seeley, 2010; Song *et al.*, 2023; Yana *et al.*, 2015) and factors that lead to design changes in construction projects have been identified (Andi *et al.*, 2007; Chang, 2002; Chang *et al.*, 2011; Cox *et al.*, 1999, Hsieh *et al.*, 2004; Love *et al.*, 2010; Perkins, 2009; Wu *et al.*, 2005). However, the aim of this section is to address the design variables from the developing country perspective.

A construction cost expert (QS) is often called upon to provide economic comparisons of alternative design variables. Building design costs are influenced by many variables, some of which are correlated (Oke *et al.*, 2023). When making a comparison, it is useful to the design team to express costs in terms of user requirements. The following are the main factors to be considered in any assessment between competing design proposals during the economic evaluation of investment in constructed facilities.

- *Site considerations*. Every construction site has its own attributes that can influence the suitability for development. Location can also influence the cost of the construction project. Site preparation is also pertinent in terms of construction costs. However, some scholars do not consider site consideration as a variable that influences construction costs (Lowe *et al.*, 2006).

- *Building size.* Building size is a variable regarding cost efficiency. This is because costs are not in proportion to changes in size. For example, factories cost more per unit than their larger counterparts.

The analysis shown in Figure 2.3 reveals that project A (15 m × 15 m × 3.5 m high) has a wall/floor ratio of 0.933, whereas the much larger project B (45 m × 45 m × 3.5 m high) has a wall/floor ratio of 0.311. This implies that cost does not change with size, but changes with the wall/floor ratio.

2.3.1 Planning efficiency

Although each outline plan may be similar in overall floor area, the way that this area is allocated within the project will be of particular importance to the client. The architect will have attempted to have made the best possible use of space within each design, but the ratio between usable and non-usable areas (circulation space) will differ. The ratio of non-usable floor area will depend on the building's function but, for planning efficiency, it should not exceed 20%. However, the planning efficiency of engineering projects such as highways will often be determined along the lines of safety considerations (Lowe *et al.*, 2006).

2.3.2 Plan shape

The plan shape of any building greatly affects costs. The more complex the shape of the building, the higher will be its cost per square metre of gross internal floor area. Generally, the simpler the geometry of the plan, the lower its unit cost. The reason why irregular shaped plans cost more can be attributed to the number of corners involved. This is known to be a factor influencing the cost of blockwork and the output of blocklayer, and it can also have some effect on roofing costs. Figure 2.4 shows an example of how building plan

Figure 2.3 Example of cost implication of building size

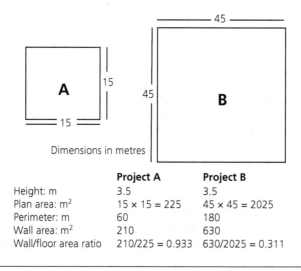

	Project A	Project B
Height: m	3.5	3.5
Plan area: m²	15 × 15 = 225	45 × 45 = 2025
Perimeter: m	60	180
Wall area: m²	210	630
Wall/floor area ratio	210/225 = 0.933	630/2025 = 0.311

Figure 2.4 Effect of building shape

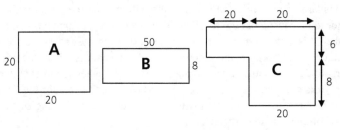

Dimensions in metres

	Project A	Project B	Project C
Height: m	3.5	3.5	3.5
Plan area: m²	400	400	400
Perimeter: m	80	116	108
Wall area: m²	280	406	378
Wall/floor area ratio	0.70	1.02	0.95

shape affects cost implications. The analysis shows that the smaller the wall to floor ratio, the more economic will be the design. As shown in Figure 2.4, a square plan shape is the most economically compact shape, with the lowest wall/floor area ratio and hence the most cost effective. However, a regular or square shape may not always provide the best economics.

2.3.3   Storey height
Storey heights are determined by the needs of building users. Storey heights can influence the costs of the vertical circulation elements (non-usable areas), especially the services. The storey height directly affects the costs of walls and partitions, together with finishes. Buildings with a high storey height will cost more per square metre of floor area than comparative accommodation with a lower storey height.

2.3.4   Building height
The costs of tall buildings are higher than low-rise structures offering the same units of accommodation. Tall buildings are generally only preferred where land is either expensive or scarce. High-rise buildings necessitate expensive foundations, structural frames, lifts, expensive fire regulations and improved services. These variables will influence construction costs.

2.3.5   Grouping of buildings
This is another factor to be considered in any assessment between competing design proposals. Grouping of buildings can influence the overall costs of projects. Building interlinking can save construction costs and in-use costs.

2.3.6   Buildability
Buildability is described as the extent to which the building design facilitates ease of construction to completion. It is necessary to note that the ease with which buildings are erected on site can influence financial savings for the contractor.

2.3.7 Constructional details

The constructional details can have an important influence on the contract period required. The use of standard or prefabricated components will shorten the construction period and will be a factor to be considered by the client. However, such components can be expensive in terms of the QS's time, and only those elements that are cost sensitive are analysed in detail. Cost studies are not normally undertaken in circumstances where the cost difference between alternatives is obviously going to make very little change to the overall cost. Where a cost study reveals only marginal savings, it would be inappropriate for the QS to lay great emphasis on such a result. In connection with new processes or technologies, the QS should remember that contractors are cost conscious.

2.3.8 Life-cycle costing

In the past, the main focus of clients was on how to reduce construction costs. However, over time and with clients becoming more enlightened, emphasis has shifted from construction costs to the 'Rs' – running, repairs and replacement costs (Ashworth and Perera, 2015). The introduction of maintenance-free construction is to minimise in-use costs.

2.3.9 Other factors that affect construction developments

2.3.9.1 Effect of government legislation on projects

Legislation is a general term used to describe the laws enacted in a country. National fiscal policies can influence construction development and the effects of government legislation on construction projects cannot be over-emphasised. For example, policies that impose stringent credit restrictions can mitigate development. In other words, the influence of government policies is pertinent to the growth of construction work. Records also show that some federal and state governments have invested money in building factories (e.g. for the cement, iron and steel industries) and have encouraged other overseas and indigenous investors (e.g. Dangote Cement). In Nigeria such ventures include Tower Aluminium, Flag Aluminium and Critical Hope General Metal Products. Legislation also affects costs due to population density (groupings of buildings on site influence construction costs) and special installations. For example, in high-rise residential blocks, a lift has to be installed in buildings above four storeys. Other special installations required by legislation that have impacts on project costs include firefighting equipment, fire and burglary alarms, refuse disposal facilities and so on.

2.3.9.2 Effect of site condition on project cost

Every construction project is unique and has its own attributes, which can influence the total cost of the development. Some of the pertinent variables are as follows.

- *Location of site*. Building costs on a site in an urban area (e.g. Abuja, Lagos, Port Harcourt, Accra, Pretoria and Johannesburg) can be 30% higher than erecting a similar building in a rural location (e.g. Igarra, Irrua and Sabo). This is due to the higher costs in urban areas (e.g. labour and materials) (Amadi, 2023). In addition, some parts of a (developing) country are subject to higher rainfall than others, which can result in greater loss of working hours.
- *Demolition and site clearance*. The degree of demolition and site clearance varies from site to site.

- *Contours.* Contours vary from one construction site to another.
- *Ground conditions.* The ground strata to establish the load bearing capacity varies from site to site, which influences the type of foundation that will be required.
- *Services.* Services can influence construction costs. These services include communication networks, gas, electricity, water supply and mains drainage.
- *Plant, materials and labour availability.* A scarcity of labour will influence construction costs. Thus, the availability of labour should be investigated, as well as the availability of materials and plant.

2.3.9.3 Effect of use of plant on construction costs

With increases in labour costs due to inflation, many housing developers/construction contractors are making greater use of plant. Studies have shown that the use of plant such as cranes enhances productivity and efficiency.

2.4 Summary

This chapter covered the background and forms of approximate estimating, the aim of which is to forecast the cost of a building project very early in the design stage. The chapter concluded with economic comparisons of alternative design variables and the effects of various factors on costs. The next chapter provides an in-depth description of cost planning, control and analysis of construction projects.

REFERENCES

Amadi A (2023) Integration in a mixed-method case study of construction phenomena: from data to theory. *Engineering, Construction and Architectural Management* **30(1)**: 210– 237, https://doi.org/10.1108/ECAM-02-2021-0111.

Andi S, Winata S and Hendarlim Y (2007) Faktor-faktor penyebab rework pada pekerjaan konstruksi. *Civil Engineering Dimension* **7(1)**: 22–29 (in Indonesian).

Ashworth A and Perera S (2015) *Cost Studies of Buildings*. Routledge, London, UK.

Ashworth A, Hogg K and Higgs C (2013) *Willis's Practice and Procedure for the Quantity Surveyor*. Wiley, London, UK.

Ball M (2014) *Rebuilding Construction: Economic Change in the British Construction Industry*. Routledge, London, UK.

Chang AST (2002) Reasons for cost and schedule increase for engineering design projects. *Journal of Management in Engineering* **18**: 29–36.

Chang AST, Shih S and Choo SY (2011) Reasons and costs for design change during production. *Journal of Engineering Design* **22(4)**: 275–289.

Cox DI, Morris PJ, Rogerson H and Jared EG (1999) A quantitative study of post contract award design changes in construction. *Construction Management and Economics* **17(4)**: 427–429

Genc O (2023) Identifying principal risk factors in Turkish construction sector according to their probability of occurrences: a relative importance index (RII) and exploratory factor analysis (EFA) approach. *International Journal of Construction Management* **23(6)**: 979–987.

Hsieh T, Lu S and Wu C (2004) Statistical analysis of causes for change orders in metropolitan public works. *International Journal of Project Management* **22(8)**: 679–686.

Love DEP, Edwards JD, Watson H and Davis P (2010) Rework in civil infrastructure projects: determination of cost predictors. *Journal of Construction Engineering and Management* **136(3):** 275–282.

Lowe DJ, Emsley MW and Harding A (2006) Relationships between total construction cost and project strategic, site related and building definition variable. *Journal of Financial Management of Property and Construction* **11(3):** 165–180, https://doi.org/10.1108/13664380680001087.

Lowe DJ, Emsley MW and Harding A (2007) Relationships between total construction cost and design related variables. *Journal of Financial Management of Property and Construction* **12(1):** 11–24, https://doi.org/10.1108/13664380780001090.

Oke EA, Adetoro EP, Stephen SS *et al.* (2023) *Risk Management Practices in Construction.* Springer Science + Business Media LLC, London, UK.

Orzeł B and Wolniak R (2022) Digitisation in the design and construction industry— remote work in the context of sustainability: a study from Poland. *Sustainability* **14(3):** 1332.

Perkins R (2009) Sources of changes in design–build contracts for a governmental owner. *Journal of Construction Engineering and Management* **135(7):** 588–593.

Sarker BR, Egbelu PJ, Liao TW and Yu J (2012) Planning and design models for construction industry: a critical survey. *Automation in Construction* **22:** 123–134.

Seeley HI (2010) *Building Economics*, 4th edn. Palgrave Macmillan, New York, NY, USA.

Song T, Pu H, Schonfeld P *et al.* (2023) Mountain railway alignment optimization integrating layouts of large-scale auxiliary construction projects. *Computer-Aided Civil and Infrastructure Engineering* **38(4):** 433–453.

Wu C, Hsieh T and Cheng W (2005) Statistical analysis of causes for design change in highway construction on Taiwan. *International Journal of Project Management* **23(7):** 554–563.

Yana AGA, Rusdhi HA and Wibowo MA (2015) Analysis of factors affecting design changes in construction project with partial least square (PLS). *Procedia Engineering* **125:** 40–45.

Chapter 3
Cost planning, control and analysis of construction projects

3.1. Design team's role in contract administration

The quantity surveying profession is evolving and in recent years its services have rapidly expanded into new areas of work. Two hundred years ago, architects were responsible not only for the design but also for the construction of works in a more direct manner and for 'cost control'. They played an important role in engineering-type projects. The overall importance of architects in the industry has diminished somewhat. Controlling cost should be complete and all-embracing if it is to contribute to stimulating confidence in the figures that construction cost consultants produce. Therefore, integrated teamwork should be encouraged. However, the efficiency of cost planning at the pre-tender stages is seriously reduced if cost control is not observed during the contract. The Royal Institute of British Architects (RIBA) prepared a suggested pattern for design team members regarding the tasks and roles expected of team members, called the 'plan of work' (RIBA, 2000). It illustrates an all-encompassing and practical analysis of the design processes, as presented in Table 3.1. Table 3.1 shows the outline, purpose of work and decision to be reached and the expected tasks of team members at each contract stage. Similarly, Figure 3.1 shows the flow of the work.

Table 3.1 Outline plan of work (continued on next page)

Stage		Purpose of work/decision to be reached	Expected tasks	Team members
A	Inception	Prepare requirement outlines	Client briefing and engage architect	Client's interests and architect as design team leader
B	Feasibility	Provide client with an appraisal and determine the form in which the project is to proceed with a focus on finance, technicality and functionality	Carry out studies of client's requirements (site conditions, planning design etc.)	Client's representative, quantity surveyor (QS), engineers and architect
C	Outline proposal	Determine approach to layout, design and construction; obtain client's approval on outline proposal	Develop the brief further and conduct studies on client's requirements	Client's representative, QS, engineers, architect and specialists, as required

Table 3.1 Continued

Stage		Purpose of work/decision to be reached	Expected tasks	Team members
D	Scheme design	Complete the brief in detail and decide on the proposal	Final brief and full design of project by the architect; the QS prepares the cost plan	Client's representative, QS, engineers, architect, specialists, all statutory and other approving authorities, as required
E	Detailed design	Obtain final decision on issues related to design, specification and construction	Full design and cost checking by collaboration of all concerned	QS, architect, engineers, specialists, and contractor (if appointed)
F	Production information	Prepare production information to enhance work decision	Preparation of specifications, schedules and drawings	QS, engineers, architect, specialists and contractor (if appointed)
G	Bill of quantities	Prepare and arrange documents for obtaining tender	Preparation of bill of quantities (BOQ) and tender documents	QS, engineers, architect, specialists and contractor (if appointed)
H	Tender action	Evaluate potential specialists and contractor via appraising	Recommendation of contractors and specialists for the work	QS, architect, engineers, contractor and client
J	Project planning	Provide planning of the site and methodology of construction activities	Establish site layout, including site handover; prepare work programme	Contractor, subcontractors, client, and architect
K	Operations on site	Provide procedure of construction activities	Establish the methodology of construction, whether manual or mechanical	QS, engineers, architect, contractor, subcontractors and client
L	Completion	Ensure that the construction project is achieved within the timeframe and cost target	Administration of building contract to practical completion	QS, engineers, architect, contractor, subcontractors and client
M	Feedback after practical completion	Analyse the management, construction and performance of the project; review of project performance in use	Building contract administration and making final inspections	QS, engineers, architect, contractor, subcontractors and client

Source: Modified from authors' work

Figure 3.1 Sequence of design team's work

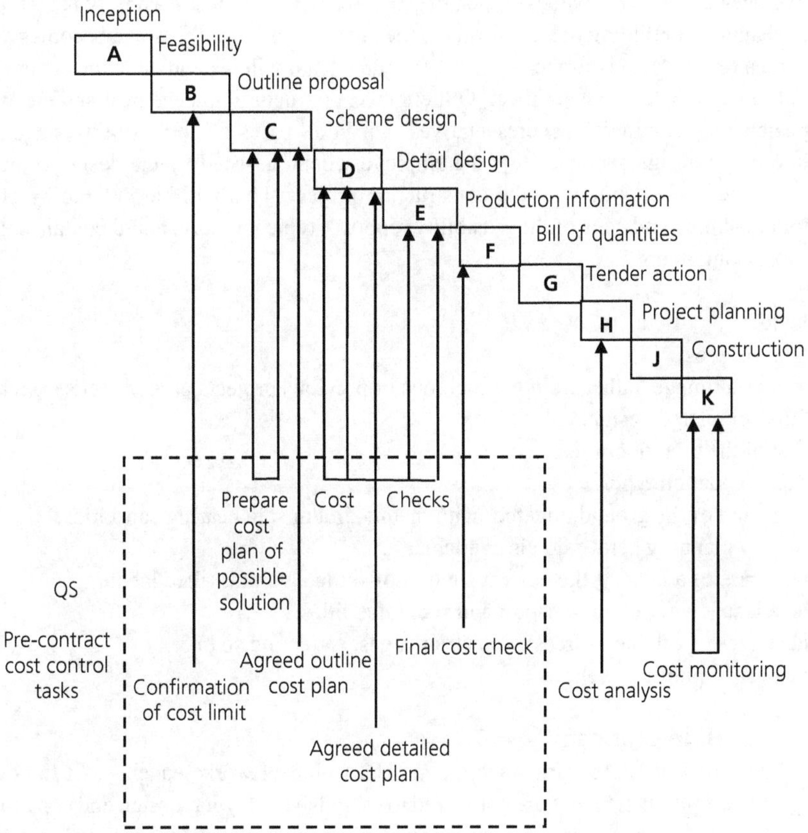

```
Inception
  [A] Feasibility
      [B] Outline proposal
          [C] Scheme design
              [D] Detail design
                  [E] Production information
                      [F] Bill of quantities
                          [G] Tender action
                              [H] Project planning
                                  [J] Construction
                                      [K]
```

QS
Pre-contract cost control tasks

- Confirmation of cost limit
- Prepare cost plan of possible solution
- Agreed outline cost plan
- Cost Checks
- Agreed detailed cost plan
- Final cost check
- Cost analysis
- Cost monitoring

3.1.1 Inception

This is the first stage in a building contract. The client/employer approaches the architect of their choice, providing them with some of the following details for guidance.

- User requirements (e.g. expected floor area).
- Budget for the project.
- The planned project's financial and time limits.
- Updates regarding the planning application.
- A list of expected design team members to engage if possible.

In the first instance, the client would wish to know the probable cost of fulfilling the requirements before continuing. Then, the client would determine how to provide for the capital outlay.

3.1.2 Feasibility

At this stage in the design sequence, the nature of the professional services of the architect will have been established, who will then be appointed to generate a feasibility report for the client. Designing a building that conforms to the client's wishes at this stage becomes a team effort, and a design team is formed. The nature of the team will depend upon the contract, but a typical lineup would be an architect, QS, engineers (structural, mechanical and electrical), and specialist subcontractor's representatives (not in all cases). The feasibility stage is the first time in the design sequence that the employer/client, assisted by the design team, will decide whether to continue with the construction project. This will be possible by closely examining the pros and cons of the feasibility report. A typical report should contain some or all the following items.

- The location of the proposed site.
- The target cost.
- A pictorial image, either from photographs of previous projects or architect's sketches, of the completed design.
- A list of the total floor areas.
- Accommodation type.
- The nature of the ground, together with the maximum safe bearing capacities.
- Whether planning permission is available.
- In the case of a factory, the nearest sources of skilled and unskilled labour.
- The adequacy of public transport and social facilities.
- The nearest available sources of electricity, gas, water and so on.
- The nearest public sewer.

3.1.3 Outline proposal

The outline proposal is the third stage of the RIBA plan of work sequence. At the outline proposal stage, options schemes are compared to establish the layout, design and construction strategy. Achieving this task involves the all-inclusiveness of the client/employer and design team members. Each team member's contribution is pertinent and can enhance the integration of decisions. For example, the architect, who is concerned with aesthetics and functional aspects, the structural engineer who is concerned with the structural forms, and the services engineer. The QS will work closely with other team members to generate the cost implications of alternative proposals as they are prepared. At this stage, the information available should have increased to include items such as outline drawings. Armed with this information, the QS prepares an outline cost plan. However, this stage is an inactive period for many design team members. The feasibility report is certainly available now, but the outline drawings are very seldom, if ever, prepared at this stage.

3.1.4 Scheme design

The architect will produce sketch plans and outline specification notes in this stage. From this, the element cost plan is prepared. The selected design cost is prepared in a statement pattern in this stage. This is the cost plan, which details the cost targets for the building elements. Pre-contract cost planning allows budgets to be assigned to the various building project elements. This provides the design team with a balanced cost framework to achieve a successful design. At this stage, the architect will have to examine the various methods of meeting the client's requirements (e.g. providing a large, uninterrupted floor area for a warehouse production line). Certainly, during this stage, the QS's advice will be sought on the cost implications of various alternative forms of construction, finishing and so on. This allows the QS to alter the cost balance within the overall target cost, as described in Section 3.1.5. According to the RIBA plan of work, the QS can accurately report to the architect at any time during the contract

- the cost of the construction project cost to the client when complete
- whether the general standards of quality being offered are too high or too low for the cost targets.

3.1.5 Detailed design

At this stage, information regarding in-depth design to enhance the final decisions regarding individual components of the building should be ready. The target cost will have been determined, with a shift from cost planning to cost controlling so that the final account does not exceed the preliminary estimate. This is also known as cost checking. Cost checking is the cost-calculating phase of the design proposals and comparing the results with the cost plan during the design process (Ashworth *et al.*, 2013). The design team members will continue to work together to secure an integrated scheme. The QS will be engaged in the preparation of cost estimates for the alternatives proposed and the QS will also ensure effective cost checking on the cost plan. This is to ensure that the client's budget is not exceeded.

Let us consider a simple hypothetical example. The target cost for the project is ₦25 000 000, with the following allocated to three elements

Element number 1 (frame)	₦8 800 000
Element number 2 (roof)	₦8 000 000
Element number 3 (services)	₦8 200 000
Total	₦25 000 000

After carefully checking the amount set for the frame with the detailed design, it is considered certain that ₦8 800 000 is an unrealistic target and at least an additional ₦1 200 000 will be needed. So, to avoid the target cost rising from ₦25 000 000 to ₦26 200 000, compensation discounts have to be made to the other elements. In this hypothetical situation, the cost target of the services was reduced: after studying the specification, a cheaper form of installation was substituted, with estimated savings of ₦1 200 000. So, the revised cost targets are

Element number 1 (frame)	₦10 000 000
Element number 2 (roof)	₦8 000 000
Element number 3 (services)	₦7 000 000
Total	₦25 000 000

Some QSs might argue that the client would prefer to pay the extra ₦1 200 000 rather than reduce the project's overall standards. The issue is the ability to establish where the line is drawn between acceptable extras of this nature and unacceptable extras and the preliminary estimate's usefulness. Does it become an outdated record? If performed correctly, there is no reason why substituting a less expensive alternative should lower standards. Each element should be cost checked after being detailed to allow reallocation of costs to be determined. After all the elements have been satisfactorily checked, the architect's drawings are passed to draughtsmen, who produce drawings of the quality necessary to produce a BOQ.

3.1.6 Production information

To enable the BOQ to be prepared, the architect's and structural engineer's offices will produce working drawings, schedules of finishes, reinforcement and so on. It is one of the functions of the QS to prepare bills of quantities. At this stage, if it becomes necessary to check the cost of any item or element, then approximate quantities should be used. This is by far the most accurate approximate estimating method. Also, establishing the preliminary tendering processes is traditional at this phase. This includes preparing advance orders for subcontractors and suppliers, the selection of subcontractors and suppliers, the form of contract, tender documents, tender invites, a questionnaire to address possible issues and a list of potential main contractors. Many of these tasks require further consultation with the design team members and the employer/client.

3.1.7 Bills of quantities (BOQs)

When the appropriate information is available, the design sequence can progress to the next stage – the preparation of BOQs. The BOQ encompasses preambles or descriptions of materials and workmanship, preliminaries and the measured works. BOQs are prepared and dispatched to the building contractors concerned. In taking off items and bill preparation by the QS, the QS ensures all details and particulars are checked, including cross-references. Where necessary, a query list for the architect and engineers covering the omission of information, ambiguities and discrepancies should be prepared. The architect and engineers can thus attend to issues demanding clarification before the tender stage. Also, the QS is responsible for ensuring the prime cost and provisional sums. The importance of BOQs cannot be over-emphasised (Ebekozien and Aigbavboa, 2024). They

- are designed and prepared as a tendering document
- provide valuable assistance with variations in pricing
- aid in computing interim valuation certificates
- offer a basis for cost planning
- can be used to identify work location if prepared in annotated format.

3.1.8 Tender action

At this stage, every task conducted is known as a tender action, which involves analysing each tender and updating information for the employer/client and design team members. Tender procedures should follow the prescribed major procedure in the Revised Public Procurement Act 2007. The core ingredients of the rules that the due process seeks to ensure compliance with are

- appropriation
- advertisement
- pre-qualification process
- pre-qualification criteria
- invitation to tender: the technical and financial bid process
- opening of tender
- the bid evaluation process
- determination of winning bid.

3.1.8.1 Appropriation

The absence of appropriation in the old procurement system gave birth to major abandonment of construction projects in Nigeria. Before 2007, Nigeria was among the few African countries with no legal framework or legislation on public procurement. For any construction project to take place, the construction project should have been appropriated in the budget. To eliminate this, the Budget Monitoring and Price Intelligence Unit (BMPIU) was established in 2001 to address inadequacies. Treasury circulars regulate BMPIU operations based on a 1958 Act that authorises the Federation Accountant General to release guidelines on public expenditure. To give legal backing to BMPIU operations, the Public Procurement Bill was sent to the National Assembly. On 31 May 2007, the Procurement Bill was passed into law and signed by President Yar'Adua GCFR on 4 June 2007. The Act put Nigeria in the league of countries with laws on how public money should be used.

3.1.8.2 Advertisement

There are two advertisement categories based on the contract's value.

- For contracts below ₦10 million, an advert must be placed on a notice board in a visible place of the parastatal, ministry, department or agency (procuring entity).
- For contracts above ₦10 million, there must be a call for pre-qualification of contractors in at least a Federal Tender Journal, government gazette or two national newspapers.

3.1.8.3 Prequalification process

The standard format set up by the due process contains

- the name and address of the procuring entity
- a brief introduction of the project description, scope, commencement time, expected completion time and so on
- the required pre-qualification entering summary
- the place and deadline for submission of applications for pre-qualification
- the availability date of pre-qualification documents.

3.1.8.4 Pre-qualification criteria

The criteria include scores up to a total of 100% based on the following.

- Evidence of business name registration or incorporation 0% (Corporate Affairs Commission).
- Registration with the Federal Ministry of Works 0%.

- Company audited accounts for the past three years 0%.
- Tax clearance certificate for the past three years 0%.
- Proof of financial capability and bank support 15%.
- Experience/technical qualifications (competence) and experience of key experts 25%.
- Similar projects executed and industry's knowledge 20%.
- Annual turnover 5%.
- VAT registration and past VAT remittances 5%.
- Bonus 3%.

The components of the bonus are

- evidence of compliance with the federal government, local content policy for building indigenous capacity 5%
- evidence of community social responsibilities.

It should be noted that, in the above list, all the parameters with zero scores are considered responsive and fundamental to contractors – the absence of any of them will inevitably disqualify the applicant.

As a pre-qualification benchmark, 70% is best practice. A contractor is deemed to be qualified and competent if the contractor scores 70% or above. This score builds confidence in the bid output. The list of all the respondents scoring 70% or above is drawn up and this list is called the list of pre-qualified bidders or list of competent bidders. All respondents scoring less than 70% are considered incompetent. Letters of situations are then issued to the successful applicants.

3.1.8.5 Invitation to tender
Tender documents containing adequate information to enable competition among the bidding contractors based on objective terms are issued to all the successful respondents who have decided to continue with the bid process. The time allowed for submission is within six weeks after the bid documents are issued.

3.1.8.6 Opening of tender
This occurs immediately after closing bidding/tendering. It takes place six weeks after the issuance of tender documents and is done in the presence of the bidders and other public members.

3.1.8.7 Bid evaluation
The tender evaluation committee constituted by the procuring entity (ministry or parastatals or agency) examines and prepares a report with award recommendations for submission to the accounting authority (permanent secretary or chief executive officer) within the procuring entity.

3.1.8.8 Determination of winning bid and validation
This is based on the recommendations of global best practices. The lowest evaluated tender cost with the requisite technical competence backed with sound financial capability is adjudged the winner. The bid submitted by the winner is termed best response bid.

After determining the winner, the procuring entity will submit a report of its procurement process handling to the BMPIU compliance review unit. If the report is deemed satisfactory, a validating certification is secured, upon which final approval for the contract award is secured.

3.1.9 Project planning

Once the client has approved the project and the letter of award has been issued to the contractor, it is necessary to prepare a programme for the whole operation from the start until, and including, commissioning. There are several useful techniques for programming, including networking, bar charts, linear programming, transportation and assignment problems, queuing theory, inventory control, simulation and scheduling. Selection of an appropriate mechanism will enhance the project to be controlled regarding time factors. The design teams and contractors should work out the appropriate planning techniques. Planning the site layout is primarily the contractor's duty, but the design team can advise if necessary.

3.1.10 Operations on site

During this stage, the contractor oversees construction activities at different stages. The QS measures work done, either by milestone or monthly as agreed. This is known as interim valuation and is forwarded to the architect for onward release of an interim certificate to the contractor and a copy is sent to the QS. This mandates the employer/client to pay within the agreed timeframe.

3.1.11 Practical completion

The contract's conditions refer to the practical completion of class 15 (Federal Ministry of Works (FMW)) work. The architect determines a date when a certificate should be issued, clearly stating that the contract has reached that stage. In determining the date of the work's practical completion, the architect should be completely satisfied with the answers to the following questions.

- Has the work been executed as per the contract document and the architect's instructions?
- Is the project in a suitable state to be taken on by the client/employer for its full and proper use?

With the release of the practical completion certificate, the following take effect automatically in accordance with the conditions of the contract.

- The contractor becomes entitled to payment of one moiety of the total of the retention fund (FMW clause 30(4)(b)).
- The defects liability period begins (FMW clause 15).
- The contractor is relieved of their obligations to ensure the works are in accordance with FMW clause 20A.
- The period of final measurement begins (FMW clause 30(5)).
- Matters for arbitration dependent upon the issuing of the certificate of practical completion can be pursued.

In some circumstances the architect may be prepared to issue a practical completion certificate and advise the contractor on specifying incomplete items. This may also apply to defective work that the architect requires to be remedied immediately, and the architect's certificate should include the schedule of uncompleted work and defective items. However, when the architect is satisfied that all defective work has been made good, the architect should issue a certificate, and is then able to consider completion of the contract. Completion is defined by the issue of the final certificate by the architect. Clause 30 requires that this certificate should be issued within three months, whichever of the following is the latest

- the end of the defect liability period or
- the completion of making good of defects or
- the receipt by the architect (or QS) of documents from the contractor relating to the nominated subcontractors and nominated suppliers' accounts.

Adjustment of the contract sum in the final account falls under the following headings.

- Variations.
- Remeasurement of provisional quantities in the BOQ.
- Nominated subcontractors' accounts.
- Nominated suppliers' accounts.
- Loss and expense caused by disturbance of regular progress of works (clause 24).
- Fluctuations in labour rates and materials prices (clause 31) (if applicable).

The QS provides the final measurement and presents the prepared document to the contractor. The QS is the sole party responsible for performing this task. In preparing the final account, the QS should provide all reasonable facilities for the contractor to be present when measurements and details are taken or recorded. Once the QS and architect have prepared the final account and are satisfied with all other points previously referred to, the architect can issue the final certificate. This certificate releases to the contractor the second moiety of the retention fund. Hence, the final certificate amount is the final account's gross amount less the amount of all previous interim payments. At this stage, the QS determines the final cost of the project. The following documents assist the QS in making decisions – agreed contractor claims, contract drawings, original priced BOQ, contract form and all recorded variations. The final account comprises adjustments for the contractor's claims (if applicable), adjustments for fluctuations, adjustments of variation account, adjustments of provisional items, adjustments of prime cost sums (if applicable), adjustments of provisional sums, the final account summary and the final account statement.

3.2. Cost planning

Cost planning is an important component of the construction management process. Construction practitioners need to understand that resources are scarce, so costs need to be managed to achieve productive outcomes. To achieve productive outcomes, stakeholders need to understand the principles of keeping the project within the budget (Ashworth and Perera, 2015; Kern and Formoso, 2006). This involves planning, estimation and control of building project costs. This came to the limelight in the 1970s and has been developed to promote accuracy in the pre-contract process. In 1972, Building Bulletin No. 4 (cited in

Bako-Biro et al. (2012)) was published, which introduced cost planning in the Department of Education and Science in the UK. This was supported by a standard list of elements from the UK Building Cost Information Service (BCIS). The BCIS was established in the 1960s with the aim of sharing information to construction-related organisations to assist their businesses' decision making (Ashworth and Perera, 2015). At the early stage of the agency, access to information was restricted to registered QSs but was later opened to firms on a subscription basis. BCIS is widely accepted for presenting cost plans in the construction industry because of the standardised approach adopted. A cost plan can be described as an estimate presented in a standard elemental format. Building a picture of planned costs over time is pertinent to cost planning. Planned costs can be unfixed or fixed. An unfixed (variable) cost occurs more than once over the life span of a construction project. A fixed cost occurs once in the life of a construction project.

Cost planning is a technique that allows every element of a proposed building project to be assigned a cost and to achieve a balanced cost framework as a guide to developing a successful design. At this stage, it is flexible and allows for budget redistribution between elements as the design progresses. RIBA formulated a procedure pattern for design team members to enhance the developing cost planning framework (taken from the *Handbook of Architectural Practice and Management* (RIBA, 2013)). A summary of Figure 3.1 and Table 3.1 is presented in Table 3.2.

Cost planning effectiveness at the pre-tender stage is curtailed if cost control is not observed for deviation from the contract. Besides determining the probable cost of a building as a key objective of cost planning and cost control, these techniques should relate cost to the design (elemental cost plan) and control the design development. Cost planning, to be at all useful, must perform certain functions. First and most important, it must provide a reliable preliminary estimate early on, usually at the feasibility stage. Without this, the time spent on producing any preliminary report becomes an academic exercise of little practical use to the client who pays for it all. Participants in the construction industry say that the first estimate of the cost presented to the client cannot be realistic, undermining the cost planning objective to produce a reliable preliminary estimate before the detailed drawings are prepared in the detailed design stage. Only this will stimulate confidence in the QS's advice. If it is not

Table 3.2 RIBA procedure pattern for design team members

Design stage	
Inception	Prepare letter of commission for the job
Feasibility	Prepare feasibility studies and determine the budget
Outline proposal	Consider alternative strategies with the client and design team and prepare cost plan
Scheme design	Carry out cost checks and update cost plan if necessary
Detailed design	Continue cost checks to ensure that the development of the design remains compatible with the cost plan
Production information	Continue cost checks on the data produced against the final cost plan

possible to provide this service, then cost planning can offer no advantages over the widely inaccurate traditional costing methods, which are so widely rejected. Cost planning creates the desires, sets out the elements and the elements' cost implications, and finally generates the project's probable cost (Ashworth and Perera, 2015). The aims of cost planning are to

- ensure that a tender for a project is within the amount budgeted by the client
- offer the employer/client value for money
- provide a balanced design solution.

3.3. Cost planning techniques

Various cost planning approaches have been identified (Ashworth and Perera, 2015; Iroegbu *et al.*, 2010; Kern and Formoso, 2006; Seeley, 2010). The application differs from project to project and there is no universal satisfactory technique for every project. There are two basic cost planning approaches in use. In practice, there is a slight variation of these methods. They are elemental cost planning and comparative cost planning.

3.3.1 Elemental cost planning

An elemental cost plan states the design team members' intentions in sums of money, which signifies the proposed construction project budget. This technique assists the design team members in developing the design within the correct economic framework via various elements of construction (BCIS standard) or constructional parts and each allocates a cost based on cost analysis of past similar projects. In this instance, costs are related to the building elements and can be reallocated (flexibly) without threats to the overall target cost. The QS (cost manager) will continue to guide the cost plan through cost checks throughout the design period. This is to ensure the cost remains within the predetermined budget (Seeley, 2010). This technique can be described as 'designing to a cost'. The technique allows cost to be presented in various ways, as presented in Table 3.3, which shows that the cost/m² of gross floor area is ₦62.6, whereas the element unit rate is ₦313/m². Detailed cost information would positively impact the element unit rate.

3.3.2 Comparative cost planning

This technique sets out the estimated cost of individual sections of work or the complete project and, where suitable, the estimated cost of alternative techniques and materials that the team may consider. The team chooses the best option with the knowledge of the cost consequences. This technique is useful in cost-in-use studies because it is based on the pricing of a design and specification. This method is also known as 'costing a design'. The method starts from the sketch plans and allows a cost study to show the various ways the design could be conducted. The cost study is based on approximate quantities. Approximate quantities

Table 3.3 Example of cost presentations

Element	Total cost of element: ₦	Cost per m² gross floor area: ₦	Element unit quantity: m²	Element unit rate: ₦/m²
Substructure	313 000	62.6	1000	313

then provide a detailed approximate estimate, which is more detailed than other methods of approximate estimating.

There is a salient difference between these major methods of cost planning. In the elemental cost planning system, the design progresses within the agreed cost limit. Regarding comparative cost planning, the design is developed at the sketch plan phase after optimal substitutes have been made.

3.4. Sources of cost information

Information is vital in construction cost decisions (Akintoye *et al.*, 1992; Ashworth and Perera, 2015) and the level of information reliability will influence confidence in subsequent decisions. The availability of cost information is essential for cost estimates prepared by QSs. The QS then communicates the construction cost information to the employer/client to ensure resourceful design, accurate pricing and building production. The desire for cost estimates and the control of construction costs gave rise to cost information management. The relevance of cost information to the building industry cannot be over-emphasised. It involves forecasting future costs, negotiating unit rates with housing developers/contractors, controlling contract prices, comparing different construction costs and preparing valuations for insurance. In collecting cost information from source(s), various factors influence the collection approach, including the timeframe, the nature of the project, the cost information available, the level of accuracy of that information, the simplicity of the information and available design information. Sources of cost data can be subdivided into the following five groups for easy reference.

- *Price books*. In Nigeria, cost information from can be obtained from published price books (e.g. the *Building and Engineering Price Book*, published by Cosines Nigeria Limited) and books published by the Nigerian Institute of Quantity Surveyors. However, in periods of high inflation, this traditional price information soon becomes outdated. In practice, price books need to be updated using indices or some subjective assessment wherever possible.
- *Cost analysis and cost models produced in-house*. These are one of the reliable sources of cost information. The data presented in these formats are generally easy to understand and interpret.
- *Building Cost Information Service (BCIS)*. This is the world's most extensive and comprehensive construction cost information service. Since its inception in the 1960s, BCIS has published cost data on various building types. It supplies extensive information on cost analyses of completed projects and offers cost indices, cost studies, cost trends and monthly briefings.
- *Priced BOQs*. These are a source of cost information, but adjusting for location is paramount. BOQs provide a wealth of information, but the prices in bills are confidential and should not be disclosed to third parties without the contractor's permission. It is also important to remember the considerable price variation that may occur between two identical bill items. The cause of this variation is described as the vagaries of tendering and can be due to many different factors.
- *Others*. Other sources of cost information are from colleagues and market surveys. Market surveys often comprise quotations from specialist subcontractors, enquiries from reputable builders' merchants, manufacturers' catalogues or quotations and trades unions agreements for labour rates.

3.5. Cost control

Regarding increasing cost and scarce resources, many construction work insist promoters on projects being designed and executed to give maximum value for money. A QS is engaged during the design stage to advise other design team members (including the architect) on the probable cost implications of their design decisions (Ashworth and Perera, 2015; Bon, 2001). As construction projects become more complex and clients more exacting in their requirements, improving and refining cost control tools becomes paramount. The cost control of a building project should start at the inception stage and end after handover to the client. Cost control is described as controlling construction project costs within the predetermined limits from inception to completion. The process of cost control commences in the strategic planning of a project. The need for vibrant cost control of construction projects cannot be over-emphasised. The relevance of effective construction cost control is as follows.

- Client requests are becoming more complicated (new techniques and materials) and more consultants are required to establish probable costs.
- These is greater urgency for building project completion.
- Employing organisations, especially the private sector, are becoming larger and insisting on innovative technologies.
- Cost control is key in eras of high inflation and high interest rates, as experienced in many developing countries such as Nigeria, Ghana and South Africa.
- The rising demand for 'green-in, green-out' and the sustainability of construction projects.
- To give clients good value for money.
- To provide a balanced design solution concerning available funds.
- To maintain expenditure within the client's construction cost budget.
- To ensure various building elements receive a balanced design expenditure.
- To achieve a balanced cost strategy for building projects considering all sustainability aspects (environmental, economic and social).
- Waste elimination/reduction and scarce global resources have stirred the need for improved mechanisms for forecasting and cost control.
- A sound cost control mechanism bridges the gap between the tender sum, the budget estimate and the final account.
- New professional software and cost tools have increased the expectation of achieving efficient building project cost control.

3.6. Definition of terms

The terminology associated with construction costs is as follows.

- *Cost planning*. This is the process of controlling construction project costs within a predetermined sum during the design stage.
- *Cost plan*. This can be described as a proposed expenditure statement on each element of a construction project.
- *Cost control*. This can be defined as all construction project cost controlling methods within the limits of a predetermined sum, from inception to completion.

- *Cost analysis*. This is a breakdown of cost data in an elemental pattern. It is used to estimate the cost limit and plan proposed construction projects.
- *Cost target*. The cost target is set against an element or work section within the cost plan.
- *Cost limit*. This is the limit of cost for an employer, beyond which the employer is not prepared to enter into an agreement.
- *Cost check*. This is the procedure of examining the estimated cost of each element as detailed designs are produced to enable the architect and other team members to compare the design cost with the cost target set against it in the cost plan.
- *Cost-in-use*. This is a method of cost prediction by which the preliminary constructional costs and the annual running and maintenance costs of a building can be abridged.
- *Life-cycle costing*. This involves assessing the initial capital costs and upcoming operating costs of a complete building over a specific time. It is used to establish effective design options.

3.7. Cost analysis

A building project cost analysis systematically documents how costs have been assigned over the various building elements. Therefore, cost analysis can be defined as the systematic breakdown of costs according to the element or work section in order to facilitate estimating the cost limit and cost planning of proposed or future construction projects. It is a process of analysing and recording the cost data of construction projects once tender information has been received. It provides information that facilitates or permits detailed comparisons between different construction projects and isolates the cause of differences in basic design. Cost analysis should be the basis of cost control (Ashworth and Perera, 2015; Okmen and Oztaş, 2010).

3.7.1 The purpose of cost analysis

The aims of cost analysis are

- to enable the design team to ascertain the total expenditure on various building elements
- to assess whether a balanced distribution of costs among the various elements is achieved
- to permit cost comparisons of the same element in different projects
- to obtain cost data for use.

A comprehensive cost analysis system has been developed in the UK (applicable in Nigeria) by the BCIS. The cost analysis is coded to identify and compare various items easily. The BCIS codes classify buildings by the form of construction, the number of storeys and the gross floor area of the building. There are four main construction types – concrete structures (C), steel-framed structures (S), blockwork construction (B) and timber-framed construction, (T) – and other forms of construction that may not fall within these classifications. For example, BCIS code B-3-200 means blockwork construction, three storeys and a $200\,m^2$ gross floor area.

The BCIS standard list of elements contains seven groups, five of which cover the building, the sixth is external works and the seventh is the preliminaries (although some say preliminaries should be disregarded, leaving six groups).

Two types of standard cost analysis are provided by BCIS – concise cost analysis and detailed cost analysis.

Concise cost analysis is a standard cost analysis form provided by BCIS, which occupies only about half a page. It provides background information about the construction project and costs of the element groupings, including the cost/m^2 of gross floor area.

The other standard cost analysis form provided by BCIS is the detailed cost analysis. This contains considerable information about the form of the construction project, design/shape information, storey height, all floor areas, floor areas suitably categorised, accommodation and design features, contract particulars, market conditions and the form of contract. The detailed cost analysis is about 3–4 pages long. The element cost is expressed in suitable units, for example

- work below lowest floor finish substructure: area of lowest floor
- upper floors: area of upper floors
- roof: area of the plan (flat) across the eaves overhang or to the inner face of the parapet wall, including roof lights
- stairs: number and total vertical rise of the stair flights or area
- external walls: area of external wall excluding windows and door openings
- windows: area of clear openings in walls
- external doors: area of clear openings in walls
- internal walls: area of internal walls excluding openings
- internal doors: area of clear openings in internal walls
- ironmongery: none
- wall finishes: area of wall finishes
- floor finishes: area of floor finishes
- ceiling finishes: area of ceiling finishes
- fittings: most times none but, where practicable, lengths or numbers of wardrobes and kitchen cabinets may be given
- plumbing installation: number and type of sanitary fittings
- electrical installation: number of power and lighting points and total electrical load
- special services: number, capacity, speed etc. for lifts, central air conditioning and so on
- drainage: none
- external works: none.

Cost equations are mathematical expressions that describe costs. In the context of cost analyses, the following are described as cost equations.

$$EC = EUQ \times EUR$$

$$EM2 = \frac{EC}{GIFA}$$

$$EM2 = \frac{ECQ \times EUP}{GIFA}$$

in which EC = element cost, EUQ = element unit quantity, EUR = element unit rate, EM2 = element cost per m², GIFA = gross internal floor area.

Quantity factors (QFs) are an expression of the quantity of a component related to the GIFA.

3.7.2 Cost analysis calculations

The cost of a proposed construction project =

$$\text{GFA (Proposed)} \times \frac{\text{Cost}}{\text{GFA (Analysed)}} \times \frac{\text{QF (Proposed)}}{\text{QF (Analysed)}}$$

where GFA = gross floor area and QF = quantity factor

Therefore the elemental cost of a proposed construction project =

$$\text{GFA (Proposed)} \times \frac{\text{Element cost (Analysed)}}{\text{GFA (Analysed)}} \times \frac{\text{Element Qty (Proposed) GFA (Proposed)}}{\text{Element Qty (Analysed) GFA (Analysed)}}$$

Also, the element cost of a proposed construction project =

Element unit cost (Analysed) × Element Qty (Proposed)

while

$$\frac{\text{Cost}}{\text{GFA (Proposed)}} = \frac{\text{Cost}}{\text{GFA (Analysed)}} \times \frac{\text{QF (Proposed)}}{\text{QF (Analysed)}}$$

Therefore,

$$\frac{\text{Cost}}{\text{GFA (Proposed)}} = \frac{\text{Element cost (Analysed)}}{\text{GFA Analysed}} \times \frac{\text{Element Qty (Proposed) GFA (Proposed)}}{\text{Element Qty (Analysed) GFA (Analysed)}}$$

Also

$$\frac{\text{Cost}}{\text{GFA (Proposed)}} = \frac{\text{Element Cost (Proposed)}}{\text{GFA (Proposed)}}$$

Table 3.4 shows a practical example.

Table 3.4 Cost analysis of proposed office accommodation for ABC Limited (BCIS code B-3-612) (continued on next page)

Element		Total element cost: ₦	Cost/GFA: ₦/m²	Element unit quantity: m²	Element unit rate: ₦/m²	Quantity factor (QF)	Brief specification
1 Substructure		650000	1062.09	150.00	4333.33	0.25	Pad and strip foundation with concrete bed on hardcore
2 Superstructure							
2A	Frame	500000	816.99	—	—	—	Reinforced concrete (RC) columns and beams
2B	Upper floor	510000	833.33	285.00	1789.47	0.47	RC suspended slab
2C	Roof	700000	1143.79	150.00	4666.67	0.25	Aluminium covering on timber trusses
2D	Stairs	250000	408.50	—	—	—	RC and timber balustrades with handrails
2E	External walls	710000	1160.13	360.00	1972.22	0.59	225 mm sandcrete walls, filled with weak concrete
2F	Windows and external doors	300000	490.20	125.00	2400.00	0.20	Aluminium doors and windows
2G	Internal walls and partitions	450000	735.29	210.00	2142.86	0.34	225 mm sandcrete walls, filled with weak concrete
2H	Internal doors	350000	571.90	81.00	4320.99	0.13	Timber panelled doors
Group element total		**3770000**	**6160.13**				

3 Finishes

3A	Wall finishes	500000	816.99	1.98	Texcoat paint on rendered wall
3B	Floor finishes	300000	490.20	1210.00	Vitrified ceramic tiles on screeded bed
				0.85	
3C	Ceiling finishes	250000	408.50	520.00	Asbestos ceiling and painted soffit
				0.25	
				150.00	
Group element total		**1050000**	**1715.69**		

4 Fittings and furnishings

		350000	571.90	Kitchen cabinet and wardrobes

5 Services

5A	Sanitary appliances	250000	408.50	WCs, sinks, basins, bidets, etc.
5B	Services equipment			
5C	Disposal installations	100000	163.40	Waste water pipes and fittings
5D	Water installation	300000	490.20	Water supply pipes and fittings
5E	Heat source	150000	245.10	Boiler installations
5F	Space heating and air treatment	100000	163.40	Air conditioners, heaters etc.
5G	Ventilating			
5H	Electrical installations	250000	408.50	Lights, accessories, cables etc.
5I	Gas installations	60000	98.04	Storage cylinder and concrete base
5J	Lift and conveyor installation			

Table 3.4 Continued

Element		Total element cost: ₦	Cost/GFA: ₦/m²	Element unit quantity: m²	Element unit rate: ₦/m²	Quantity factor (QF)	Brief specification
5K	Protective installation						
5L	Communication installations						
5M	Special installations						
5N	Builder's work in connection with services	140 000	228.76				
5O	Builders profit and attendance in connection with service	160 000	261.44				Builders' associated works
Group element total		**1 510 000**	**2467.34**				
6 External works							
6A	Site works	320 000	522.88				Landscaping and grassing
6B	Drainage	120 000	196.08				Septic tank, soakaway
6C	External services	80 000	130.72				Connection of services
6D	Minor building works	110 000	179.74				Gate houses, gen house
Group element total		**630 000**	**1029.42**				
Subtotal							
7 Preliminaries		1 200 000	1960.78				
Total (less contingencies)		**9 160 000**	**14 967.35**				

Cost planning, control and analysis of construction projects

Table 3.5 Preparation of cost plan (budget estimate) based on analysed building project (office accommodation for ABC Limited) (continued on next page)

		Budget cost: ₦
Substructure	Proposed element quantity: 24 m × 12 m = 288 m² (₦4333.33/m²) = ₦1 247 999.04	
	Add say 20% for increased number of storeys = ₦249 599.81	
	Total	1 497 598.85
Frames	Analysis QF = 2/3 = 0.67	
	Proposed QF = 3/4 = 0.75	
	$\dfrac{\text{Total element cost (analysed)}}{\text{Total GFA (Analysed)}} = \dfrac{9\,160\,000}{14\,967.35} = 612.00$	
	Budget cost $= \dfrac{612 \times 816.99 \times 0.75}{0.67}$	559 699.12
Upper floors	Budget cost $= \dfrac{612 \times 833.33 \times 0.75}{0.67}$	570 893.24
Roof	Budget cost $= \dfrac{12 \times 1143.79 \times 0.22}{0.25}$	615 999.54
Stairs	Number of analysed flights = 4	
	Number of proposed flights = 8	
	Budget Cost $= \dfrac{250\,000 \times 8}{4}$	500 000.00
External walls	Budget cost $= \dfrac{612 \times 1160.13 \times 0.32}{0.59}$	385 084.51
Windows and external doors	Budget cost $= \dfrac{612 \times 490.20 \times 0.32}{0.20}$	480 003.84
Internal walls and partitions	Budget cost $= \dfrac{612 \times 735.29 \times 0.41}{0.34}$	184 498.97
Internal doors	Budget cost $= \dfrac{612 \times 571.90 \times 0.30}{0.13}$	807 698.77
Wall finishes	Area of walls = (167.1 + 281.7) × 2 = 897.6 m²	
	Proposed QF $= \dfrac{897.6}{612} = 1.47$	
	Budget cost $= \dfrac{612 \times 816.99 \times 1.47}{1.98}$	371 210.55
Floor finishes	Budget cost $= \dfrac{612 \times 490.20 \times 1}{0.85}$	352 944
Ceiling finishes	Budget cost $= \dfrac{612 \times 408.50 \times 0.21}{0.25}$	210 001.68
Fittings and furnishings	Number of analysed floors = 3	

Table 3.5 Continued

		Budget cost: ₦
	Number of proposed floors = 4	
	Budget cost = $\dfrac{350\,000 \times 4}{3}$	466 666.67
Services	As applicable to fittings and furnishings	2 013 333.33
	Budget cost = $\dfrac{1\,570\,000 \times 4}{3}$	
External works	Cost of external works for analysed project is like the proposed work	630 000
Preliminaries	Experience has shown that factors affecting preliminaries in the analysed project are similar to the proposed project except for extensive use of some mechanical plants, such as crane and hoist due to the height of the proposed building, which may increase the cost of preliminaries (1 200 000) by 10%	1 320 000
	Sum	10 965 633.07
	Add design and price risk, say 2%	219 312.66
	Total	**11 184 945.73**
	Add price levelling factor or increase in price between the time an estimate is prepared and tender for the proposed project Analysed index = 125 Proposed index = 128	
	Price level = $\dfrac{128-125}{125} \times 100 = 2.4\%$	268.438.70
	Add contingency, say 10%	1 145 338.44
	Estimated total cost	**12 598 722.87**

3.8. Summary

This chapter discussed the design team's role during the contract administration stages (pre- and post-contract in line with the RIBA plan of work (RIBA, 2000). At the tender stage, the chapter covered the Public Procurement Act 2007 regarding due process. This includes the appropriation, advertisement, pre-qualification process and criteria, tender invitation and opening, evaluation and determination of winning of bid. Also covered in this chapter were cost planning, cost analysis and cost control. The chapter concluded with the purpose of cost analysis. This includes enabling the team to ascertain the total expenditure on various building elements, assess the cost distribution among the elements, compare the same element's costs in different buildings and generate cost data for construction project planning. The next chapter gives an in-depth description of the nature of the cost and cost-in-use, including the various techniques.

REFERENCES

Akintoye SA, Ajewole O and Olomolaiye OP (1992) Construction cost information management in Nigeria. *Construction Management and Economics* **10(2)**: 107–116, https://doi.org/10.1080/01446199200000011.

Ashworth A and Perera S (2015) *Cost Studies of Buildings*. Routledge, London, UK.

Ashworth A, Hogg K and Higgs C (2013) *Willis's Practice and Procedure for the Quantity Surveyor*. Wiley, London, UK.

Bako-Biro Z, Clements-Croome DJ, Kochhar N, Awbi HB and Williams MJ (2012) Ventilation rates in schools and pupils' performance. *Building and Environment* **48**: 215–223.

Bon R (2001) The future of building economics: a note. *Construction Management & Economics* **19(3)**: 255–258, https://doi.org/10.1080/01446190010020354.

Ebekozien A and Aigbavboa C (2024) *Professional Practice for Quantity Surveyors in the 21st Century*. Taylor & Francis, London, UK.

Iroegbu A, Nwafo K, Wogu C and Ogba SI (2010) Application of project cost planning techniques in the Nigeria project cost management system. *The Coconut: A Multidisciplinary Journal of Environment, Agriculture, Science and Technology* **2(1)**: 1–8.

Kern AP and Formoso CT (2006) A model for integrating cost management and production planning and control in construction. *Journal of Financial Management of Property and Construction* **11(2)**: 75–90, https://doi.org/10.1108/13664380680001081.

Okmen O and Oztaş A (2010) Construction cost analysis under uncertainty with correlated cost risk analysis model. *Construction Management and Economics* **28(2)**: 203–212.

RIBA (Royal Institute of British Architects) (2000) *Outline Plan of Work*. RIBA, London, UK.

RIBA (2013) *Handbook of Architectural Practice and Management* RIBA, London, UK.

Seeley HI (2010) *Building Economics*, 4th edn. Palgrave Macmillan, New York, NY, USA.

Andrew Ebekozien and Clinton Aigbavboa
ISBN 978-1-83549-841-5
https://doi.org/10.1108/978-1-83549-838-520241004
Emerald Publishing Limited: All rights reserved

Chapter 4
Life-cycle costing for building projects

4.1. Introduction to life-cycle costing

There is increasing inflation and high interest rates in many developing countries, and there is thus a call for mechanisms to reduce construction project costs. Making this decision involves selecting substitutes to achieve the project objectives. This scenario promotes the life-cycle costing (LCC) concept and its application in construction projects. In an era of increasing inflation and high interest rates, eliminating waste and reducing costs is the call of the hour, hence the need to embrace the LCC concept (Ashworth and Perera, 2015; Goh, 2016; Goh and Sun, 2015; ISO, 2001; Opawole et al., 2020). LCC has been used in advanced and some developing countries (Opawole et al., 2020). Clients in advanced countries are more knowledgeable regarding the implications of the whole-life cost of a construction project. LCC is a tool in construction cost studies for evaluating the whole cost of a project over time, so the assessment includes the costs associated with the project from construction to disposal. The LCC concept is most effective with options for consideration if conducted at the early stages of inception/briefing, feasibility and conceptual design. The birth of LCC or 'whole-life costing' as a terminology makes 'costs-in-use' obsolete (Ferry et al., 1999).

In building projects, key stakeholders' perceptions regarding LCC are relative. Yaman and Tas (2007) stated that end users, investors (financiers), design team members and clients are concerned with project costs in different ways due to their background, expectations and goals. The issues of including upcoming operational and maintenance costs at the design phase are complex. Oduyemi (2015) identified weaknesses in LCC for determining the maintenance and performance of building elements, especially in projects in developing countries. This adds to the complexity. In many developing countries, using Nigeria as a case study, the adoption of LCC techniques has not been fully embraced in cost estimations of projects (Opawole et al., 2020), and conventional methods still dominate cost estimations of building projects in Nigeria. The aim of this chapter is to identify the encumbrances facing the use of LCC techniques in building projects in developing countries, using Nigeria as a case study. The identified encumbrances will enlighten construction practitioners regarding areas that should be paid attention in order to mitigate future hindrances and improve the implementation of LCC techniques. The outcomes will enhance the practical implementation of LCC, achieve value for money and improve the cost performance of projects.

4.2. Fundamentals of life-cycle costing

Every construction project life cycle originates from inception, to post-practical completion, maintenance and ending with final disposal. LCC can be described as a financial estimate of a proposed construction project, including the project's initial, operational, maintenance and disposal costs over its operational life. The International Organization for Standardization (ISO, 2001) defines LCC as

> an economic assessment considering all agreed projected significant and relevant cost flows throughout analysis expressed in monetary value. The projected costs are those needed to achieve defined performance levels, including reliability, safety, and availability.

According to Langdon (2007), Goh (2016) and Opawole *et al.* (2020), LCC is a tool for calculating building whole-life project cost (including initial, operating, maintenance and disposal costs). LCC can be used to evaluate diverse options, with a focus on achieving the client's objectives. The concept is a process that has been developed within the built environment sector. Ferry and Flanagan (1991: p. 9) define the process as

> …putting the estimated capital, maintenance, operating and replacement costs into a comparable form and bringing them into a single figure which allows for the fact that these items of expenditure will take place at different stages within the timescale.

Besides LCC as a decision-making and contract management tool, it can be used to predict the total costs of a building project's life and then select alternative solutions. This is one of LCC's novelties and is significant in the construction industry, especially regarding building costs and offering clients (public and private sectors) details regarding the expected sum to be received when projects degenerate (Oduyemi, 2015). Therefore, LCC is a tool that offers information to analyse the economic feasibility of construction projects regarding cost-effectiveness, cost drivers and appraisals of many building options.

In the building industry, many LCC procedures have been generated for project productivity and efficiency. Langdon (2007), Olubodun *et al.* (2010), Goh (2016) and Opawole *et al.* (2020) identified many processes for LCC implementation. Langdon (2007) stated that the iterative features of LCC make it unique – proposals and measures/answers are provided, assessed and tested at the inception phase before integrating them. To achieve this progression, many decisions are involved, regarding construction materials, construction methods, sustainability and other factors. The iteration process offers the certainty of life cost as it develops from inception to demolition.

4.3. Conducting a life-cycle cost analysis

In many developing countries, LCC tools have lax usage and varying opinions of LCC tools, despite increased public–private partnership routes and private finance initiatives. LCC application in project cost estimating has received less empirical evaluation. In the Nigerian construction industry, LCC application evidence was, in 2010, restricted to the environmental sustainability of industrial production process and energy use evaluation (Opawole *et al.*, 2020). This also applies to, for example, Togo, Benin Republic, Liberia, and Ghana. In many African countries, LCC usage was restricted to sectors other than the built environment sector (Dunmade, 2019). In Ghana, the focus was on fossil energy consumption cost implications and environmental impact analysis (Afrane and Ntiamoah, 2012) and LCC empirical application in environmental impact analysis (Rivela *et al.*, 2014). In Nigeria, extant studies on LCC usage focus on environmental implications and the cost of waste management (Ogundipe and

Jimoh, 2015), restricted LCC analyses of industrial production procedures (Olaniran *et al.*, 2017), the cost estimating process and the energy efficiency of buildings (Kwag *et al.*, 2019), environmental impacts, costs and performance of substitute power system analysis (Salisu *et al.*, 2019 and the barriers to LCC application in building projects (Opawole *et al.*, 2020).

Opawole *et al.* (2020) noted 14 LCC tools used in Nigeria, including life-cycle sustainability evaluation tools for assessing social, economic and environmental factors, life-cycle assessments, design for recycling and design for remanufacturing, modularity, materials and disassembly. Dunmade (2019) clustered the tools into life-cycle allied processes, life-cycle evaluation tools and design tools. Training via upskilling and reskilling regarding the application of LCC tools must be balanced in the era of construction digitalisation and smart project execution. In many developing countries, Nigeria and Ghana included, the reality is that practitioners need to catch up on the LCC tools that enhance building cost estimation and contract management decision-making.

4.4. Encumbrances to the application of life-cycle costing in construction

Stakeholders in the construction industry, especially in developing countries, are concerned that the use of LCC is still low, despite the advantages linked with the use of LCC for estimating building costs (Dunmade, 2019; Opawole *et al.*, 2020). In 2010, Seeley (2010) noted that LCC knowledge and its usage was in its infancy. The wide gap between industry and academia complicates the situation. Various issues hindering LCC application in the built environment are summarised in Table 4.1. Opawole *et al.* (2020) clustered the encumbrances into four main areas – (*a*) procurement policy issues, (*b*) reliable information and data availability, (*c*) client interest, cost and communication effectiveness and (*d*) professional skills and motivation challenges.

Table 4.1 Encumbrances to LCC adoption in building projects (modified from Opawole *et al.* (2020: p. 507)) (continued on next page)

Identified encumbrance
Inadequate expertise of LCC professionals
Lack of awareness of LCC benefits and uses
Inadequate clarity of LCC principles
Lax interest from practitioners
Lengthy payback period (return on investment) for building projects
Impact of inflation on forecasted figures
Inadequate LCC knowledge
Inadequate quality of data
Absence of a standardised approach
Client reluctance
Project teams' attitudes to LCC tools
Separation of capital and running costs of buildings

Table 4.1 Continued

Identified encumbrance
Absence of reliable data
Type of investor
Absence of common standards
Risk and uncertainty
Impact of experience on the quality of planning
Inadequate readily accessible and reliable information and guidance
Uncertainty and risk tolerance
Data collection is costly and time-consuming
Financial hindrances
Short-term budgeting by client
Cost of undertaking the exercise
Hindrances with satisfying multiple institutional stakeholders
Not needed by clients
Lack of interest in LCC implications
Unstable economic conditions
Absence of fiscal encouragement
Unreliable data, involving several assumptions
Ineffective communication among team members
Unwillingness of professionals to develop decision-making methods
Low interest from clients
Inadequate monitoring
Lax requests for sustainable materials and products
Lax government policies
Fragmented nature of the industry
Poor maintenance culture
Unwillingness to change management policies and strategies
Dealing with intangibles
Overabundance of cost models
Impact of out-of-plan future results
Inadequate procurement award motivations
Restrictions to sustainable options usage
Market conditions and assumptions
Bureaucratic structures in administrative procedures
Restrictions from public procurement legislation
Fragmented nature of project teams

4.5. Life-cycle costing methods
4.5.1 Traditional LCC
Some scholars (Ferry *et al.*, 1999; Flanagan and Norman, 1984; Goh and Sun, 2015; Opawole *et al.*, 2020) support the use of a discounted present value method for making economic assessments of relevant costs. Besides the support that started when the LCC concept was established, the extant literature on LCC focuses on the discounted present value technique, describing it as a conventional approach.

4.5.2 Improved traditional LCC
To improve the conventional method, Flanagan *et al.* (1987) proposed a risk management system in LCC to account for the risk and uncertainty associated with the future, affirming that LCC deals with an unknown future. The assumptions used to underline the cost-estimating exercise must therefore be balanced. Flanagan *et al.* (1987) noted that, by applying probability and sensitivity studies to LCC, the emerging findings would quantitatively show the influence of assumptions on decisions. The essence of LCC is to meet clients' demands, looking for value for their investment (Goh and Sun, 2015).

4.5.3 Non-traditional LCC
Over the years, the conventional method has progressed into mathematical model formulations to estimate LCC parameters (Kirkham *et al.*, 2002; Kishk, 2004). Kirkham *et al.* (2002) used stochastic modelling to attain theoretical probability density functions and analyse the LCC of a case study. Kishk (2004) established a mechanism that adds significant data and subjective data to find innovative measures to represent uncertainty issues in LCC modelling. A mathematical framework enhances the integration of random and non-random data (Goh, 2016; Goh and Sun, 2015).

4.5.4 Green building LCC
Studies conducted in early 2000 regarding LCC application methods revealed supportive signs of the operating advantages of 'green' construction, including the design. It was shown that green buildings have low occupancy costs because of the lower consumption of utilities (e.g. power, water) and reduced maintenance costs. Aye *et al.* (2000) noted that, in making the best sustainable development selection, the LCC and advantages derived from it should be considered and assessed in monetary terms for each option by employing the conventional discounted present value technique to create respective net present values for comparison (Goh and Sun, 2015).

4.6. Tools and techniques for life-cycle costing (investment appraisal)
These can be divided into two – discounting methods and conventional methods.

4.6.1 Discounting methods
4.6.1.1 Net present value (NPV)
To determine the NPV of an investment, the estimated net-of-tax cash flow is discounted to the tune of the initial capital outlay, and the value of initial capital is set at a level that would give the shareholder a rate of return at least equal to what they could obtain outside the

company. For a given cash profile for project x over a planning horizon of n years (for $t = 0, 1, 2, ..., n$) and a given value of the minimum attractive rate of return (MARR) = 1, the NPV is given by:

$$NPVx = \sum_{t=0}^{n} At \times (ITI) - t = \sum_{t=0}^{n} At \times (P/F.i.n.t)$$

where At = initial capital, ITI = initial capital set at a level, t = time, P = profit, i = index and F = future.

The net future value at $t = n$ is given by

$$NPVx = \sum_{t=0}^{n} At \times (ITI) - t = \sum_{t=0}^{n} At \times (P/F.i.n.t)$$

It can easily be shown that:

$NFVx = NPVx\,(1+i) - n = NPVx\,(F/P, i, n)$ or

$NPVx = NFVx\,(1+i) - n = NFVx\,(P/F, i, n)$

where NFV = net future value.

Consequently, if $NPVx \geq 0$, it follows that $NFVx \geq 0$ and vice versa. It is neutral when the value is 0 and may be treated as the acceptable limiting condition for the project.

Worked example 1

Ehi-Ose & Partners desire to invest in Auchi Polytechnic, Edo State, Nigeria. Two options are open to the firm. Option A is a shopping plaza and Option B is a multi-purpose centre encompassing cinema halls, conference halls and fitness centres. Table 4.2 shows the cash flow for the options. The task is to determine which project should be selected.

The MARR is set at 10%. Unless stated otherwise, all sums of currency are in Nigerian naira (₦).

Table 4.2 Cash flow for Option A (shopping plaza) and Option B (multi-purpose centre)

	Option A	Option B
Annual gross income	10 500 000	18 000 000
Residual	70 000 000	90 000 000
Planning horizon	20 years	20 years
Initial cost	95 000 000	150 000 000
Annual maintenance	500 000	5 000 000

Solution 1
Option A

$$(\text{NPV}_1)\ 10\% = 95 + (10.5 - 0.5)(\text{PIU},\ 10\%,\ 20) + 70(\text{PIF},\ 10\%,\ 20)$$
$$= 95 + 10(\text{PIU},\ 10\%,\ 20) + 70(\text{PIF},\ 10\%,\ 20)$$
$$= 95 + 10(8.5136) + 70(0.1486)$$
$$= 0.538$$

Option B

$$(\text{NPV}_2)\ 10\% = 120 + (18 - 5)(\text{PIU},\ 10\%,\ 20) + 90(\text{PIF},\ 10\%,\ 20)$$
$$= 120 + 13(\text{PIU},\ 10\%,\ 20) + 90(\text{PIF},\ 10\%,\ 20)$$
$$= 120 + 13(8.5136) + 90(0.1486)$$
$$= 4.0508$$

Here, PIU is the profitability index unit and and PIF is the profitability index future.

Option B has the higher NPV and hence should be selected.

The NFV can also be computed for both options, using the MARR as the reinvestment rate for the cash flow of the project before the end of the planning horizon (F = future, P = profit, I = index and U = unit).

$$(\text{NFV}_1)\ 10\% = -95 + (F/P,\ 10\%,\ 20) + (10.5 - 0.5)(F/U,\ 10\%,\ 20) + 70$$
$$= -95(F/P,\ 10\%,\ 20) + 10(F/U,\ 10\%,\ 20) + 70$$
$$= -95(6.7275) + 10(57.2750) + 70$$
$$= 3.6375$$

$$(\text{NFV}_2)\ 10\% = -120 + (F/P,\ 10\%,\ 20) + (18 - 5)(F/U,\ 10\%,\ 20) + 90$$
$$= -120(6.7275) + 13(57.2750) + 90$$
$$= 27.275$$

Note the same result is obtained.

$$\text{NFV}_1 = \text{NPV}_1(F/P,\ 10\%,\ 20)$$
$$= 0.538(6.7275)$$
$$= 3.619$$

$$\text{NFV}_2 = \text{NPV}_2(F/P,\ 10\%,\ 20)$$
$$= 4.050(6.7275)$$
$$= 27.252$$

Note that the answer would be the same if these figures were approximated to one decimal place.

Worked example 2

Bowen Partners, an estate developer, approved building a luxury hostel block for students of the Faculty of Environmental Sciences at the University of Benin, Edo State. The construction period is estimated to be two calendar years. The firm has agreed with AB Bank to pay ₦25 million to the contractor every month. At the end of construction, Bowen Partners will pay back the principal and interest to the bank and take possession of the property. The annual net rental income from the hostel is expected to be ₦9.25 million for the next 50 years, after which the University of Benin will take over ownership of the hostel. If the MARR of the firm is 12%, the task is to determine if the construction project is worthwhile.

Solution 2

Cost of construction plus interest at construction (i.e. year 2)

$$Q = 2.5(F/U, 2\%, 24) \text{ (i.e. 24 months of bank payment @ 2\% per month)}$$
$$= 2.5(30.4218)$$
$$= 76.0545$$

At the end of construction, the owner invests this amount (i.e. ₦76.0545)

$$\text{NPV} = -76.0545 + 9.25(P/U, 12\%, 50)$$
$$= -76.0545 + 9.25(8.3045)$$
$$= 0.762125$$
$$= 762.13k$$

The NFV method can be used for three conditions to compare project worthiness

- the options compared have equal economic lives
- the options compared have unequal economic lives
- all options compared have unlimited economic lives.

When all options being considered have equal economic lives, there is a challenge among the options. The calculation is relatively straightforward, with the option delivery of the highest NPV determined to be the most favourable. When the options being compared have unequal economic lives, the NPV computation must be manipulated to compare the options over the same number of years. This can be achieved by comparing the options over a period equal to the least common multiple of their economic lives. This method is particularly used to evaluate physical assets that are assumed to last for a long time (e.g. buildings, ports, dams).

4.6.1.2 Internal rate of return (IRR)

This is the most common discounting method of investment appraisal. It can be defined as the rate of interest that, when used to discount the net-of-tax cash flows of a proposed investment, reduces the net present value (NPV) of the project to zero. This discount rate can be found by trial and error; if a negative NPV results, the rate chosen is too high; if a positive NPV is obtained, the rate is too low. Although it appears to involve many calculations, it should only be necessary to carry out up to two trial discounts, the true IRR then being determined by interpolation. However, the IRR does not take into consideration the reinvestment opportunities related to the timing and intensity of the outlays and returns at the intermediate points over the planning horizon.

The IRR is the value for r, which satisfies:

$$\sum_{t=0}^{U} \frac{At}{(1+i)^t} = 0$$

Worked example 3
For a construction project with cash flow spreading only within two periods, the IRR can be calculated as shown in Table 4.3.

Table 4.3 IRR calculation for a construction project with cash flow spreading only within two periods

Year	Cash flow
0	−500
1	+750

Solution 3

$$\begin{aligned}
\text{NPV} &= -500 + 750(1+i)^{-1} \\
0 &= -500 + 750(1+i)^{-1} \\
0 &= -500(1+i) + 750 \\
500(1+i) &= 750 \\
(1+i) &= 750/500 \\
(1+i) &= 1.5 \\
i &= 1.5 - 1 \\
i &= 0.5 \\
i &= 50\%
\end{aligned}$$

Worked example 4
Ebeks Associates, an estate developer, is planning a major investment in real estate near Auchi Polytechnic main gate. Ebeks Associates is ready to expend a whopping ₦450 million in developing a world-class multi-purpose complex. Annual maintenance and operation costs are estimated at ₦3 million and annual rental is estimated at a conservative ₦20 million.

The market forecast has shown that the prepared value of the property in 10 years will be ₦600 million. If the planning horizon is 10 years, calculate the IRR of this development.

Solution 4

$$NPV = -450 + (20 - 3)(P/U, i, 10) + 600\,(P/F, i, 10)$$

$$NPV = -450 + 17(P/U, i, 10) + 600\,(P/F, i, 10)$$

Arbitrarily select several values of i to calculate NPVs. Then select two i values at which the NPV changes sign.

i

$0 = -450 + 17(P/U, 0, 10) + 600(P/F, 0, 10) = 150$
$1 = -450 + 17(P/U, 1, 10) + 600(P/F, 1, 10) = 254.1921$
$2 = -450 + 17(P/U, 2, 10) + 600(P/F, 2, 10) = 194.8842$
$3 = \ldots\ldots$
$4 = \ldots\ldots$
$5 = \ldots\ldots$
$6 = -450 + 17(P/U, 6, 10) + 600(P/F, 6, 10) = 10.1617$
$7 = -450 + 17(P/U, 7, 10) + 600(P/F, 7, 10) = -25.6188$

$$\begin{aligned}
\text{IRR}\,(i) &= 6\% + 10.1617/[10.1617-(-25.6188)] \\
&= 6\% + 0.284 \\
&= 6.284\%
\end{aligned}$$

Generally, linear interpolation is best done with the two successive values of i (interest rate), where the NPV shows a sign reversal. In the case above, it was between 6% and 7%. The quantitative formula is:

$$\text{IRR} = \text{LDR} + \left(\frac{\text{LRNPV}}{\text{LRNPV} - \text{HRNPV}} \times (\text{HDR} - \text{LDR})\right)$$

where

LDR = low discount rate
LRNPV = low-rate NPV
HRNPV = high-rate NPV
HDR = high discount rate

4.6.1.3 Necessity postponability
This criterion is negative. The basis is, the more postponable an investment, the less attractive it seems. Thus, the requirement for urgency is the basis of investment decision-making. If a project could only be conducted now but could be initiated later, it would be chosen in favour of a project that could only be undertaken in the future.

4.6.2 Conventional methods
4.6.2.1 Payback method
This is a very trusted method and its name nearly describes its operation. It is the crudest form of investment criterion, but is nevertheless widely used. The payback period is the period for an investment to create incremental cash to recover the initial capital outlay in full. A cut-off point can be selected, beyond which the project will be rejected if the investment has not been paid off.

Worked example 5
Table 4.4 shows the cash flow of an investment project. Determine if the investment is acceptable, taking the payback period as 3 years.

Table 4.4 Cash flow for an investment project

	Year	Cash flow: ₦
Initial investment	0	10 000
Net cash flow	1	4000
Net cash flow	2	4000
Net cash flow	3	3000
Net cash flow	4	2000
Residual value		2000

Solution 5
Year 1 $= -10\,000 + 4000 = 6000$
Year 2 $= -6000 + 4000 = 2000$
Year 3 $= -2000 + 3000 = +1000$
$2000/3000 = 2/3 = 66.67\%$
Payback period $= 2\tfrac{2}{3}$ years

The project is thus acceptable since the payback period is within the specified 3 years.

Worked example 6
Consider the two projects shown in Table 4.5. If the projects are mutually exclusive, determine which should be selected if the specified payback period is 4 years.

Table 4.5 Cash flow for projects A and B

	Year	Net cash flow: ₦	
		Project A	Project B
Initial cost	0	−12000	−10000
Cash flow	1	5000	3000
Cash flow	2	4000	3000
Cash flow	3	3000	3000
Cash flow	4	2000	3000
Cash flow	5	1000	3000
Residual		2000	1800

Solution 6

Year	Project A	Project B
1	−12000 + 5000 = 7000	−10000 + 3000 = 7000
2	−7000 + 4000 = 3000	−7000 + 3000 = 4000
3	−3000 + 3000 = 0	−4000 + 3000 = 1000
4		−1000 + 3000 = +2000
		1000/3000 = 1/3 = 33.33% = 3⅓ years

Project A pays back within 3 years whereas project B pays back in 3⅓ years. Project A is thus the preferred project.

4.6.2.2 Average rate of return (ARR) method

The average rate of return, also called the accounting rate of return, is the ratio of profit (net of depreciation) to capital. The first decision that must be made is how to define profit and capital. Profit can be taken as either gross of tax or net of tax, but since businesses are mostly interested in their post-tax position, net profit is a more useful yardstick. Net profit can be either the profit made in the first year or the average of what is made over the project's entire lifetime. This method does not account for the incidence of cash flows so that projects with the same capital costs, expected length of life and total profitability would be ranked equally acceptable.

Worked example 7

Table 4.6 shows the cash flow of a proposal by ABC Limited to acquire a block making machine. Using the straight line method of depreciation, a working capital of ₦4000 and a 5-year economic life, evaluate using the ARR technique.

Table 4.6 Cash flow of proposal by ABC Limited to acquire a block making machine

Year	Net cash flow: ₦
0	20 000
1	8000
2	8000
3	5000
4	5000
5	3000
6 (salvage value)	2000

Solution 7

Profit can be calculated as equal to cash flow minus depreciation. The annual depreciation would be

$$\frac{20\,000 - 2000}{5} = \frac{18\,000}{5} = ₦3600 / \text{year}$$

Year 1	$+8000 - 3600 = 4400$
Year 2	$+8000 - 3600 = 4400$
Year 3	$+5000 - 3600 = 2400$
Year 4	$+5000 - 3600 = 1400$
Year 5	$+3000 - 3600 = -600$
Total profit	$= 12\,000$
Average annual profit	$= 12\,000/5 = 2400$

Initial capital employed = Initial expenditure + Working capital

$= 20\,000 + 4000 = 24\,000$

Average capital employed

$$= \frac{\text{Capital expenditure} - \text{Salvage value} + \text{Salvage value} + \text{Working capital}}{2}$$

$$\frac{20\,000 - 2000 + 2000 + 4000}{2} = 12\,000$$

Return on capital employed

$$= \frac{\text{Average annual profit}}{\text{Initial capital employed}}$$

$= 2400/12\,000 = 0.2 = 20\%$

Return on average capital employed

$$= \frac{\text{Average annual profit}}{\text{Average capital employed}}$$

$$= 2400/15\,000 = 0.16 = 16\%$$

4.6.2.3 Optimal investment criterion

Some investment cash flows have more than one sign reversal during their economic life or time horizon. In such cases, there is the likelihood of more than one value of IRR. This makes the IRR method immeasurable in such a project evaluation. The external rate of return (ERR) is the discount rate used for reinvestment and finance. The ERR is used where cash flow is positive and is the continuous return possible from the available funds under the existing market.

Therefore, the historical ERR (HERR) is one technique that allows unconventional cash flow problems with non-unique IRRs to be solved. The HERR is the interest rate that sets the future value (FV) of the project net benefit or profit compound at the available reinvesting rate with expenses or costs.

Worked example 8

Comparing the HERR of the investment project depicted in Table 4.7, an assumed opportunity exists in the financial market to reinvest positive cash flow at 8.0%.

Table 4.7 HERR of an investment project

Year	Cash flow
0	−200
1	100
2	150
3	−150
4	80
5	60
6	+50

Solution 8

For HERR

$$\text{NFV} = -200(F/P, i, 6) + 100(F/P, 8.0\%, 5) + 150(F/P, 8.0\%, 4) - 150(F/P, i, 3) + 80(F/P, 8.0\%, 2) + 60(F/P, 8.0\%, 1) + 50$$

$$\text{NFV} = -200(F/P, 8.0\%, 5) + 150(F/P, i, 3) + 100(F/P, 8.0\%, 5) - 150(F/P, 8\%, 4) + 80(F/P, 8.0\%, 2) + 60(F/P, 8.0\%, 1) + 50$$

$$\text{NFV} = -200(F/P, i, 6) - 150(F/P, i, 3) + 100(1.4693) + 150(1.3605) + 80(1.1664) + 60(1.0800)$$

$$\text{NFV} = -200(F/P, i, 6) - 150(F/P, 3) + 50 + 509.117$$

$$\text{NFV} = -200(F/P, i, 6) - 150(F/P, i, 3) + 559.117$$

5%: $-200(1.3401) - 150(1.1576) + 559.117 = 117.457$

6%: $-200(1.4185) - 150(1.1910) + 559.117 = 101.777$

10%: $-200(1.7716) - 150(1.3310) + 559.117 = 5.147$

11%: $-200(1.8704) - 150(1.3867) + 559.117 = -20.103$

Then

$$\text{HERR} = 10\% + \left(\frac{5.147}{5.147 - 20.103}(11-10)\%\right)$$
$$= 10\% + 0.2038\%$$
$$= 10.2038\%$$

4.7. Summary

This chapter discussed life-cycle costing (LCC) as a concept to add value and eliminate waste in the whole life cycle of a construction project. Many stakeholders are interested in a project's initial cost without considering operational and maintenance costs. The barriers to the use of LCC techniques in building projects in developing countries were identified, using Nigeria as an example. Methods used for investment appraisal were also identified and examples were given. The next chapter gives an in-depth description of cost indices, including the factors to consider when constructing an index.

REFERENCES

Afrane G and Ntiamoah A (2012) Analysis of the life-cycle costs and environmental impacts of cooking fuels used in Ghana. *Applied Energy* **98**: 301–306.

Ashworth A and Perera S (2015) *Cost Studies of Buildings*. Routledge, London, UK.

Aye L, Bamford N, Charters B, and Robinson J (2000) Environmentally sustainable development: a life-cycle costing approach for a commercial office building in Melbourne, Australia. *Construction Management and Economics* **18(8)**: 927–934.

Dunmade I (2019) Industrial applications of lifecycle tools in West Africa: an overview of trends. *Procedia Manufacturing* **35**: 1141–1145.

Ferry DJO and Flanagan R (1991) *Life Cycle Costing – A Radical Approach*. Construction Industry Research and Information Association, London, UK, CIRIA Report 122.

Ferry DJO, Brandon P and Ferry J (1999) *Cost Planning of Buildings*, 7th edn. Blackwell, Oxford, UK.

Flanagan R and Norman G (1984) *Life Cycle Costing for Construction*. Surveyors Publications, London, UK.

Flanagan R, Kendell A, Norman G and Robinson GD (1987) Life cycle costing and risk management. *Construction Management and Economics* **5(4)**: S53–S71.

Goh HB (2016) Designing a whole-life building cost index in Singapore. *Built Environment Project and Asset Management* **6(2)**: 1–20, https://doi.org/10.1108/BEPAM-09-2014-0045.

Goh HB and Sun Y (2015) The development of life-cycle costing for buildings. *Building Research & Information* **44(3)**: 319–333, https://doi.org/10.1080/09613218.2014.993566.

ISO (International Organization for Standardization) (2001) ISO 12006-2: Building construction. Organization of information about construction works. Part 2: Framework for classification of information. ISO, Geneva, Switzerland.

Kirkham RJ, Boussabaine AH and Awwad BH (2002) Probability distributions of facilities management costs for whole life cycle costing in acute care NHS hospital buildings. *Construction Management and Economics* **20(3)**: 251–261.

Kishk M (2004) Combining various facets of uncertainty in whole-life cost modelling. *Construction Management and Economics* **22(4)**: 429–435.

Kwag BC, Adamu BM and Krarti M (2019) Analysis of high-energy performance residences in Nigeria. *Energy Efficiency* **12**: 681–695.

Langdon D (2007) *Life Cycle Costing as a Contribution to Sustainable Construction: Guidance to the Use of the LCC Methodology and its Application in Public Procurement. Final Report.* See http//ec.europa.eu/enterprise/sectors/construction/!les/compet/life_cycle_costing (accessed 10/10/2022).

Oduyemi OI (2015) *Life Cycle Costing Methodology for Sustainable Commercial Office Building.* University of Derby, Derby, UK.

Ogundipe FO and Jimoh OD (2015) Life cycle assessment of municipal solid waste management in Minna, Niger State, Nigeria. *International Journal of Environmental Research* **9(4)**: 1305–1314, https://doi.org/10.22059/IJER.2015.1022.

Olaniran JA, Jekayinfa SO and Agbarha HA (2017) Life cycle assessment of cassava flour production: a case study in Southwest Nigeria. *Journal of Engineering and Technology Research* **9(1)**: 6–13, https://doi.org/10.5897/JETR2015.0580.

Olubodun F, Kangwa J, Oladapo A and Thompson J (2010) An appraisal of the level of application of life cycle costing within the construction industry in the UK. *Structural Survey* **28(4)**: 254–265.

Opawole A, Babatunde SO, Kajimo-Shakantu K and Ateji OA (2020) Analysis of barriers to the application of life cycle costing in building projects in developing countries: a case of Nigeria. *Smart and Sustainable Built Environment* **9(4)**: 503–521, https://doi.org/10.1108/SASBE-11-2018-0057.

Rivela B, Moreira MT, Bornhardt C and Feijoo G (2014) Life cycle assessment as a tool for the environmental improvement of the tannery industry in developing countries. *Environmental Science and Technology* **38(6)**: 1901–1909, https://doi.org/10.1021/es034316t.

Salisu L, Enaburekhan JS and Adamu AA (2019) Techno-economic and life cycle analysis of energy generation using concentrated solar power (CSP) technology in Sokoto State, Nigeria. *Journal of Applied Science and Environmental Management* **23(4)**: 775–782, https://doi.org/10.4314/jasem.v23i5.1.

Seeley HI (2010) *Building Economics*, 4th edn. Palgrave Macmillan, New York, NY, USA.

Yaman H and Tas E (2007) A building cost estimation model based on functional elements. *A|Z ITU Journal of the Faculty of Architecture* **4(1)**: 73–87.

Andrew Ebekozien and Clinton Aigbavboa
ISBN 978-1-83549-841-5
https://doi.org/10.1108/978-1-83549-838-520241005
Emerald Publishing Limited: All rights reserved

Chapter 5
Cost indices in the construction industry

5.1. Introduction

The role of cost indices is pertinent in collating construction cost data and forecasting the probable cost of projects in order to achieve reliable building project costs. The use of cost indices enhances financial commitments and budgeting at the early stage (Amadi, 2023). Building costs are established by materials, plant, labour costs, overheads and profits. These variables prices vary regarding substitute components or over time (Gichunge et al., 2010; Hassanein and Khalil, 2006a, 2006b; Seeley, 2010). Gichunge et al. (2010) noted that building function cost indices can evaluate tender differences. Cost index measures or cost-related changes in an item or items from one location are linked with time to another (Olagunju et al., 2014). Many items are weighted according to their significance within the index. It is important that a base date is selected and given the value of 100, and all future values/figures either decrease or increase. Cost index interpretation is understood better when its purpose is explicit. Ssegawa (2003), Hassanein and Khalil (2006a, 2006b) and Gichunge et al. (2010) reported that many organisations in developing countries do not publish construction data and such absence of cost data can stall the planning and pricing of building projects. In Nigeria, there is a challenge with reliable construction cost indices, especially regarding estimating the future cost trends in the sector. Unstable inflation and economic policies have not helped matters.

5.2. Relevance of cost and price indices in a developing country's construction industry

- Cost and price indices can be employed to update elemental cost analyses of a building project. This process allows tender information from past projects to be upgraded to present costs for estimating. The formula for updating is:

$$\text{Proposed rate} = \text{Original rate} \times \left(\frac{\text{Current index}}{\text{Original index}}\right)$$

Thus, the percentage difference between the two rates can be calculated as:

$$\text{Percentage change} = \left(\frac{\text{Current index} - \text{Original index}}{\text{Original index}}\right) \times 100$$

- Cost and price indices can be used to update construction data for cost research purposes. The outcomes make it easy to analyse the cost trends and patterns and a researcher can extrapolate a trend into the future with a reliable prediction. An example is the tender price index (Akintoye et al., 1998; Hassanein and Khalil, 2006a).
- Regarding calculating building price fluctuations, the Joint Contract Tribunal provides these in the contract.
- Regarding building cost relationships, indices based on different construction materials are germane. For example, concrete frames can be compared with steel frames. The findings from such a comparison can enhance decision-making to identify which option is preferable. Building cost indices are useful in life-cycle costing techniques. They compare alternative best measures to a design issue (Masu, 1987).
- For forecasting, cost indices are key to assessing overall market conditions with updated prices and reliable forecasts for the future.

5.3. Types of cost indices

The construction industry publishes several cost indices that guide practitioners regarding changes in building costs/prices. Tysoe (1981), Ferry and Brandon (1991), Gichunge et al. (2010) and Goh (2016) identified two main types of cost indexes – factor cost indices and tender price indices, as presented in Table 5.1.

5.3.1 Tender price indices

These cost indices are popular with quantity surveyors (QSs) because the document is prepared by the Building Cost Information Service (BCIS) (BCIS, 1993; Seeley, 2010). The generated index compares the present figure with the base year total (Gichunge et al., 2010).

Table 5.1 Cost indices for the building industry

	Type	Content
Building Cost Information Service (BCIS) building and element price indices	Factor cost indices (allows for market conditions adjustment)	Combined index: building and engineering services
BCIS building and services cost indices	Factor cost indices	Building (inclusive of all materials – concrete, brickwork/blockwork, steel frame) and engineering services
BCIS tender price indices	Tender price indices	Building
Building housing cost index	Factor cost index	Housing (especially residential)
Spon's building cost index	Factor cost index	Building
Department of Education construction output price indices	Output price indices	Building
Department of Building and Engineering (DB & E) tender price	Tender price index	Housing

Source: Authors' work

In principle, a tender price index is developed by the analysed bills of quantities for tenders and comparing the bill rates against a base schedule of rates. In many developed countries (e.g. Singapore), a national agency provides stakeholders with a platform for tender price analysis (Goh, 2016). This is not the situation in many developing countries. Nigeria has lax and insufficient industry cost information about building activities and maintenance. An agency should coordinate and streamline cost data to improve this situation. Setting up a national integrated database of Nigeria's building data sources and the engagement of private stakeholders are possible ways of moving forward. To strengthen this move, there should be an institutional framework to coordinate data from Nigerian contractors. Such a database is important for contractors to share information and reference for research purposes.

The quality of construction data is also critical: it should be transparent, reliable, valid and in adherence to globally agreed standards. It is important to express non-disclosure agreements to enable employers and their design team members to release contract details such as rates, elemental costs, operational costs and other associated costs. When sourcing construction cost data from past building projects, this is a challenging issue. Seeley (2010) highlighted some advantages of a tender price index, including the following.

- The building cost is well represented to the employer/client against the contractor's cost.
- It is done independently, rather than those that rely on the summations of others.
- It offers a tendering climate suggestion.
- It assists in cost-planning effectiveness.
- Reliable cost limits can emerge and be applied.
- Price determinants can be evaluated with ease.
- It can be used to update historical cost data.
- It shows the industry's inflation trend with an emphasis on capturing relative price change.
- It can be used for forecasting market conditions.

However, practitioners also acknowledge some challenges inherent in using tender price indices, the most frequent of which is the difference between the final project cost and the tender price. Many factors contribute. The contract may contain fluctuation clauses for materials and labour. Also, the provisional and prime cost sums may have been approximated. In some cases, the unit rates in the approved bills of quantities may not be realistic. Hence, the fundamentals of how the index was established should be understood if it is to be used wisely. In summary, tender price indices are attempts to analyse the inflation of the contract prices between housing developers/contractors and employers/clients for developing new building projects (Marco and Graham, 2008).

Economic indicators are key variables that can influence tender price indices. From the UK perspective, Akintoye (1991) identified 23 indicators that influence tender price indices. In Hong Kong, the economic indicators influencing tender price indices were identified as the employment rate for forecasting the directional of changes, money supply, implicit gross domestic product (GDP) deflation, the GDP of the construction industry, GDP, consumer price indices, building cost indices and the best lending rate (Ng *et al.*, 2000). These indicators were clustered into construction-related, banking sector-related and economic-related indicators (Ng *et al.*, 2004). In other studies, the unemployment rate, money supply, implicit

GDP deflator, GDP in construction, GDP, composite consumer price indices, building cost indices and bank interest rate were identified as significant economic indicators (Wong and Ng, 2010). Of the eight indicators, construction GDP, GDP and building cost indices are significant in influencing the prediction of tender price indices. In most developing countries, including Ghana, the construction industry is related to the country's economy since the government is the largest employer in construction (Chileshe and Yirenkyi-Fianko 2012; Ebekozien *et al.*, 2023a, 2023b). There is a positive relationship between the construction industry and the economic indicators (consumer price index, producer price index, interest rate, exchange rate, GDP etc. (Osei, 2013)). Nketiah and Obeng-Aboagye (2016) reported that the composite consumer price index, GDP and interest rate influence the Ghanaian economy. Ernest *et al.* (2019) evaluated the 23 indicators identified by Akintoye (1991) and other extant literature and found that the composite consumer price index, currency exchange rate, interest rate, producer price index and GDP were the most significant economic indicators that can influence tender price indices.

5.3.2 Factor cost indices or building cost indices

With these indices, the cost of each constituent resource (e.g. materials, plants, labour) is assessed. An index weighting is conducted to indicate each constituent in line with the total building value. The method uses plant and equipment input costs, materials and labour changes. For example, when estimating the cost of reinforced concrete, allowance is made for cement, sand, coarse aggregate, formwork, steel and other items regarding the material costs percentages and the various categories of workers needed. Allowance is also made for overheads and profit. Many factors influence the percentage that allows for the overhead, including the market conditions, the economic situation, the client type and the type of building project (Gichunge *et al.*, 2010). In the application of factor cost indices to adjust a cost analysis rate or a tender to allow for time variations, the formula is:

$$\text{Percentage change} = \left(\frac{\text{Index figure at adjusted date} - \text{Index figure at present}}{\text{Index figure at present}} \right) \times 100$$

Gichunge *et al.* (2010) identified various factor indices.

- Factor indices for each building constituent (including material, equipment and labour). The inability to consider the tender market is a main disadvantage.
- The Laspeyres index is used where an index applies a base year weighting. This is common due to the challenge of creating current quantity values for others (Yu and Ive, 2008). The formula is:

$$\left(\frac{\text{Current price} \times \text{Base quantity}}{\text{Base quantity} \times \text{Base price}} \right) \times 100$$

- The Paasche index is applied where the index uses weightings obtained from the present year under consideration. It uses current-year quantities in its computation:

$$\left(\frac{\text{Current price} \times \text{Current quantity}}{\text{Base quantity} \times \text{Base price}} \right) \times 100$$

- Fisher's ideal index uses the average of Paasche and Laspeyres indices to analyse the project/constituent.

5.3.3 Other published indices

Ashworth (2010) identified other published indices, described in the following sections.

5.3.3.1 BCIS building cost indices

The BCIS indices monitor equipment/plant, materials and labour costs of housing developers/building contractors. The National Economic Development Office (NEDO) formula indices are used to compile the work. A weighting system is obtained from analysed approved tenders. The following is an example.

- Index 2 – general cost index. This comprises all construction works but exempts services engineering works.
- Index 3 – general cost index. This is based on a weighting of indices 4–6: blockwork (index 4) 50%, steel frame (index 5) 25%, concrete frame (index 6) 25%.
- Index 4 – blockwork construction cost index (comprises all categories of residential building types).
- Index 5 – steel-framed construction cost index.
- Index 6 – concrete-framed construction cost index.

5.3.3.2 BCIS tender price index

The BCIS tender price index has similar features to the Department of Education (DoE) tender index. This type of index is limited to government building projects.

5.3.3.3 Building indicators

The economic performance of the building industry is illustrated via building magazines. To achieve this goal, critical indicators or areas are examined. Regarding key indicators or areas, these are relative and varied.

5.3.3.4 Housing and construction statistics (DoE)

The Directorate of Statistics in the DoE manages this quarterly publication, with a focus on indicators (including housing finance, house building performance, architects' workload, starts and completion of dwellings, new orders etc.).

5.3.3.5 Davis Langdon cost forecasts

The chartered QSs in this firm regularly prepare cost forecasts from their past building projects and publish them in the *Architects' Journal*. The costs are limited to London and its environs. The published cost forecasts also appear in the *Builder's Price Book*, which is a reliable cost book and reference materials for the building industry.

5.3.3.6 Nationwide Building Society Index

The focus of this index is on different house prices and their change over time to guide intending purchasers. Data from these indices are documented on a regional basis. Similar cost data are collected and published by the Royal Institution of Chartered Surveyors (RICS) for their subscribers and members. For over three decades, the RICS has established a costs database for the building industry (Goh, 2016). The affiliates in most developing countries, such as the Nigerian Institute of Quantity Surveyors, have more work ahead of them.

5.3.3.7 The National Economic Development Office (NEDO) method
NEDO has researched several methods in an attempt to devise a formula to measure construction cost increases over a certain duration and compare different materials. Experts in the UK have embraced this approach for calculating increased costs in preference to the conventional method.

5.4. The hedonic regression method
This technique can be employed to develop building price indices. This approach considers the attributes of construction materials and can monitor price changes. To achieve this goal, two models can be adopted – the explicit time-variable model and the strictly cross-sectional model (Olagunju et al., 2014).

5.4.1 Explicit time-variable hedonic index
This hedonic model comprises discrete periods as autonomous constructs and grouped data for adjacent periods. Hence, a single regression is used to estimate the coefficients of a time dummy and implicit prices of housing variables. The estimation is achieved by regressing the natural logarithm of the constant selling price and the duration, including time dummies and several observed independent variables (Olagunju et al., 2014).

5.4.2 Strictly cross-sectional hedonic index
This is a substitute for the explicit time-variable model that assists in controlling potential changes to the implicit house price factors over a duration. This model allows implicit characteristic prices to vary over time. The functional form is like the explicit time-variable hedonic model, but the time dummies are excluded from the equation (Olagunju et al., 2014).

5.5. Challenges in choosing cost indices
Deciding which price or cost index to employ in a building project is challenging for the QS. Besides an error that could influence the quality and reliability of the evaluation/analysis, the high inflation and unstable economic policies in many developing countries have not helped matters. Seeley (2010) highlighted the variables that should be reflected in the industry's cost indices, including

- the work's variety and the need for diverse industries
- the updating of contract tenders, evaluating changes in housing developers' costs
- many and diverse challenges facing cost assessments
- the importance of integrating various series of indices to cover differences in time points.

The diversity of construction work influences the challenges facing price measurements or basic costs. This extends to new construction projects because of their uniqueness from start to finish. Costs are computed at each construction stage. Tysoe (1991) and Seeley (2010) suggested the factors that should be considered when selecting cost indices for building projects, including

- the aim of the index
- item selection

- the choice of weighting
- the choice of the base year.

Ashworth (2010) identified hindrances facing the usage of cost indices in the building industry.

- The index composition is based on commodities that should measure change. Items to measures may be unrepresentative.
- Assumed commodities considered relevant could be outside the index scope.
- The same source, quantity and item for the commodity should be used to measure contrasts, but this may be impossible.
- Alternative items/commodities as substitutes for the original may increase inaccuracies.
- Any slight changes in fashion may counter the correct items previously chosen. Change is inevitable over time.
- Individual weightings can influence the index's significance for certain items/commodities.
- Errors in computation due to false returns can trigger inaccuracies in the data.

5.6. Summary

The chapter covered the principles of cost indices in developing countries, using Nigeria as an example. Cost indices shows how building costs have changed over time. The relevance of cost indices in the building industry was discussed and major cost indices were identified. The factors to consider when constructing an index were also described, including index purpose, item selection, choice of weights and choice of base year. The chapter concluded with the issues that make the use cost/price indices difficult. The next chapter gives an in-depth description of property valuation and development budget.

REFERENCES

Akintoye SA (1991) *Construction Tender Price Indices: Modelling and Forecasting Trends.* PhD thesis, University of Salford, Salford, UK.

Akintoye A, Bowen P and Hardcastle C (1998) Macro-economic indicators of construction contract prices. *Construction Management and Economics* **16**: 159–175.

Amadi AI (2023) A weather-adaptive perspective to building cost estimation in Nigeria: exploring climate-induced cost variance. *International Journal of Building Pathology and Adaptation*, https://doi.org/10.1108/IJBPA-01-2023-0004.

Ashworth A (2010) *Cost Studies of Buildings*, 5th edn. Pearson Education, Harlow, UK.

BCIS (Building Cost Information Service) (1993) *Quarterly Review of Building Prices.* BCIS, London, UK.

Chileshe N and Yirenkyi-Fianko BA (2012) An evaluation of risk factors impacting construction projects in Ghana. *Journal of Engineering Design and Technology* **10(3)**: 306–329.

Ebekozien A, Aigbavboa C, Samsurijan MS et al. (2023a) Smart contract applications in the built environment: how prepared are the Nigerian construction stakeholders? *Frontiers of Engineering Management* **11**: 50–61, https://doi.org/10.1007/s42524-023-0275-z.

Ebekozien A, Abdul-Aziz, AR and Jaafar M (2023b) Mitigating high development and construction costs of low-cost housing: findings from an empirical investigation. *International Journal of Construction Management* **23(3)**: 472–483, https://doi.org/10.1080/15623599.2021.1889748.

Ernest K, Theophilus AK, Amoah P and Emmanuel BB (2019) Identifying key economic indicators influencing tender price index prediction in the building industry: a case study of Ghana. *International Journal of Construction Management* **19(2)**: 106–112.

Ferry DJ and Brandon P (1991) *Cost Planning of Buildings*, 6th edn. Granada Publishing, London, UK.

Gichunge HK, Masu SM and K'Akumu OA (2010) Factor cost indices practices in the building industry for Nairobi, Kenya. *Journal of Financial Management of Property and Construction* **15(1)**: 61–70, https://doi.org/10.1108/13664381011027980.

Goh BH (2016) Designing a whole-life building cost index in Singapore. *Built Environment Project and Asset Management* **6(2)**: 1–20, https://doi.org/10.1108/BEPAM-09-2014-0045.

Hassanein AAG and Khalil BNL (2006a) Building Egypt 1 – A general indicator cost index for the Egyptian construction industry. *Engineering, Construction and Architectural Management* **13(5)**: 463–480, https://doi.org/10.1108/09699980610690747.

Hassanein AAG and Khalil BNL (2006b) Developing general indicator cost indices for the Egyptian construction industry. *Journal of Financial Management of Property and Construction* **11(3)**: 181–194, https://doi.org/10.1108/13664380680001088.

Marco KW and Graham I (2008) The compilation methods of building price indices in Britain: a critical review. *Construction Management and Economics* **26(7)**: 693–705, https://doi.org/10.1080/01446190802043918.

Masu SM (1987) *Application of Life Cycle Costing Technique to the Construction Industry in Kenya. A Case Study of Nairobi Kenya Railways Buildings Nairobi*. MA thesis University of Nairobi, Nairobi, Kenya.

Ng ST, Cheung SO, Skitmore K, Lam KC and Wong LY (2000) Prediction of tender price indices directional changes. *Construction Management and Economics* **18(7)**: 843–852.

Ng ST, Cheung SO, Skitmore M and Wong TC (2004) An integrated regression analysis and time series model for construction tender price indices forecasting. *Construction Management and Economics* **22(5)**: 483–493.

Nketiah NK and Obeng-Aboagye S (2016) *Economic and Financial Indicators – A Tool for Investment Decisions*. See. http://www.omegacapital.com.gh/articles/ECONOMIC%20INDICATORS.pdf (accessed 24/12/2016).

Olagunju M, Owolabi KM and Afolayan AO (2014) Determination of a price index for escalation of building material cost in Nigeria. *International Journal of Managerial Studies and Research* **2(10)**: 116–133.

Osei V (2013) The construction industry and its linkages to the Ghanaian economy – polices to improve the sector's performance. *International Journal of Development and Economic Sustainability* **1(1)**: 56–72.

Seeley HI (2010) *Building Economics*, 4th edn. Palgrave Macmillan, New York, NY, USA.

Ssegawa JK (2003) A building price index: A case for Botswana. *Journal of Financial Management of Property and Construction* **8(1)**: 1–12.

Tysoe BA (1981) *Construction Cost and Price Indices*. E & FN Spon, London, UK.

Tysoe BA (1991) Making the right choice. *Chartered Quantity Surveyor* November, pp. 12–13.

Wong JM and Ng ST (2010) Forecasting construction tender price indices in Hong Kong using vector error correction model. *Construction Management and Economics* **28(12)**: 1255–1268.

Yu MKW and Ive G (2008) The compilation methods of building price indices in Britain: a critical review. *Construction Management and Economics* **26**: 693–705.

Andrew Ebekozien and Clinton Aigbavboa
ISBN 978-1-83549-841-5
https://doi.org/10.1108/978-1-83549-838-520241006
Emerald Publishing Limited: All rights reserved

Chapter 6
Property valuation and developer's budget

6.1. Methods of determining the value of property

This section discusses the various approaches to determining property value in developing countries, using Nigeria as a case study. Property valuation is a way of determining the financial value of a property. The goal may be for a mortgage (sale purposes) or determining a property tax bill (tax purposes), which is important for concerned stakeholders such as insurance firms, local tax authorities, lenders, investors and so on. Property values are determined by an appraisal, and many factors influence the value of a property. Maleta (2012) identified the cost of development, government assistance, legal considerations, planning control requirements, available utility services, ground conditions, building size and shape, location and development type as factors that could affect a property valuation. Besides these factors, the building's internal attributes (conventional or smart construction, aesthetics, quality of construction, total square areas, number of rooms etc.), property demand level, the economic situation, government policy towards housing provision and neighbourhood characteristics (safety and distance to urban area) are key factors that influence the value of property. Three methods for determining the value of a property are as follows.

- *Sales comparison method*. This is the most common approach of the known methods. This method involves an appraiser comparing a property to a nearby property to guide the selling price.
- *Income method*. The method is common with investment properties because of their income streams. In this method, an appraiser will assess all necessary factors (e.g. market condition, vacancy rates, rental rates, expense statements) to determine the income to be generated from the property. Income and cash flow forecasts are used to establish the property's value.
- *Cost method*. This method allows the value of a property to be evaluated based on its replacement cost (i.e. the cost to reproduce or replace the current building on the same land). When the replacement cost is established, depreciation is deducted and the site value is added to determine the current property value.

6.2. Investment

Construction investment is one of the key components of gross domestic product, and it is capital-intensive. This section focuses on property investment. Property investment has long-term socio-economic, political, environmental and security consequences if not well-regulated and managed (Fadun and Saka, 2018; Taylor *et al.*, 2023). Regulating property investment is key to controlling economic policy, as revealed in the housing markets of some developed countries like the USA (Jagun, 2020) and the UK (Taylor *et al.*, 2023). Thus, property investors should be regularly upskilled and reskilled to enhance their investment decisions. It is an ever-growing market, even in developed countries. The House of Commons Library (HCL, 2018) and the Ministry of Housing, Communities and Local Government (MHCLG, 2019) reported that an estimated 340 000 new houses are needed every year in England to bridge the housing demand–supply gap. Recent issues associated with many property investments have raised concern among stakeholders. Various risk factors may have triggered these issues, including unrealistic interest rates, changes in development demand, inability to meet client objectives and so on. Archibong and Ogunba (2018) and Jagun (2020) argued that some of these factors are poorly articulated in the preparation of feasibility and viability appraisals. Integrating these risk factors is pertinent because most reports address key outcomes based on an understanding of present market situations.

In Nigeria, the concept of property investment appraisal started gaining attention in the 1970s because of the crude oil boom. Investors started showing interest in investing in Nigerian sectors, including the property market, because of this boom (Ajayi, 1996; Olaleye *et al.*, 2007). In the mid-1980s, Nigeria's property market faced challenges because of the crude oil crises and global recession. This resulted in many failed property development investments. Despite global economic recovery, Nigeria's economy and property market continued to experience unfavourable challenges. Several factors could have contributed, including high unemployment, economic instability, high interest rates, absence of sustainable government housing policies, high construction costs, insecurity, kidnapping and banditry, youth restiveness, foreign exchange rates (naira losing value) and high inflation. Making informed decisions regarding property investment in undiversified economies such as Nigeria is thus important. However, certain risk factors should be considered in the process of decision-making regarding investments and property development in any economy. These include:

- fundamental economic uncertainty
- timing
- options over risk.

It is pertinent to report these risk factors to improve development values and return on investment (Farayibi, 2005; Jagun, 2020), and any report should be all-inclusive (covering, for example, feasibility, commencement of the property's construction to completion, commission and facility management). Each stage has its challenges and associated risk factors. Several models can be used for the property development process. The model proposed by Wiegelmann (2012) is adopted in this chapter, which is a four-tier framework comprising

- project initiation (client's development concept)
- conception (feasibility analysis)
- management (costs, time and quality)
- marketing/disposal (letting/sale).

The model emphasises that the development process is not significant because of the evolving character of the property development market. Haliza-Asat *et al.* (2017) validated the significance of key development process phases.

Identifying risk in investment appraisal is key to mitigating the quantifiable loss that could emanate from foreseeable investment returns. This is because of known catastrophes and economic inflation (Hallegatte *et al.*, 2016). Investment risk exists because stakeholders are unable to predict uncertainties in prospective future events perfectly. Many factors can hinder cash flow projections for property development, including unpredictable future returns, national government policies and political issues. The research of Gupta and Tiwari (2016) is one of the few studies identifying the critical factors influencing investments in property development using a hypothetical methodology. Gupta and Tiwari (2016) found errors ascribed to environmental, technological, physical and human uncertainty that affect the market inadequacy. Property development risks are classified into market, interest rate, business, financial, political, taxation, tenant, legal, relative, timing, planning, management and union risks.

The pre-construction stage of property development in developing countries has been considered (Gupta and Tiwari, 2016; Haliza-Asat *et al.*, 2017; Jagun, 2020; Newell and Razali, 2009; Nnamani, 2018; Ogunbayo *et al.*, 2018). In Nigeria, Nnamani (2018) assessed the application of quantitative techniques for risk analysis in the investment appraisal of properties. Ogunbayo *et al.* (2018) explored the application of risk analysis by stakeholders during pre-development evaluation in Nigeria. Nnamani (2018) discovered that use of the independent assessment rate is significantly high and also revealed that lack of technical ability for software usage, unreliable data sources, high complexity and restricted knowledge are the main issues hindering the application of quantitative risk analysis techniques in Nigeria. Ogunbayo *et al.* (2018) reported that conventional appraisal practices cannot account for uncertainty and risk in property development appraisal. Risk management in property appraisal cannot be over-emphasised.

Risk management in property development is not new but implementation is lax in many developing countries, including Nigeria. The concept offers the platform to quantitatively and qualitatively identify risks linked with property development (Bartelink *et al.*, 2015). It also assists guiding decision-makers in managing risk processes from an integrated perspective. Wiegelmann (2012) argued that risk occurrence in the development process should not be under-rated because it could negatively influence all facets of managing projects. Khumpaisal *et al.* (2010) suggested a risk model to structure the decision-making process for investors across the various development phases. This model implies that the mission of risk management is to devise a reference basis that allows the investor to identify, assess and manage risks across all the development stages.

6.3. Methods of valuation

In practice, estate valuers are often confronted with a series of valuation problems. The valuation of land and property is usually undertaken by an estate surveyor for a variety of different purposes – sale, purchase, occupation or investment, auction reserves, mortgage loans, inheritance tax, income tax or local taxation purposes. Property values vary considerably from one district to another, so a valuer should have extensive experience of values in the practice area. As the main valuation methods in practice, Marshall and Kennedy (1993) identified appraisals of long-term funding arrangements, forward funding appraisals, ground rental valuations, sensitivity analysis, discounted cash flow appraisal, cash flow approach, the feasibility or residual profit method and the residual method of valuation (conventional method). Each of these techniques has merits and disadvantages. They all use estimated or known variables to compute the unknown variable. The approach applies to land value, profit or the rate of return.

6.3.1 Comparison method

This popular valuation technique compares a property directly with prices paid in the open market for similar properties, where reasonably close substitutes are available, and transactions occur frequently. Its prime use is for residential properties where there is likely to be a greater similarity between different properties.

6.3.2 Contractor's method

The basis of the contractor's method is that the value of a property is equivalent to the cost of erecting the building together with the cost of the site. However, this is an unsound assumption as the value of a property is determined not by what it costs to build but by the amount that purchasers in the open market are prepared to pay for it and the price the seller is prepared to take. Its main use is in connection with valuations for insurance purposes and for buildings such as schools, churches, hospitals and so on, for which there may be little in the way of comparative valuations.

6.3.3 Residual method

This method is commonly used to value land to be developed or redeveloped with entirely new buildings or land with existing buildings that are to be refurbished. This method works on the premise that the price that a purchaser can pay for such land is the surplus after taking out the proceeds from the sale of the furnished development, the costs of construction, the costs of purchase and sale, the cost of finance and an allowance for profits required to carry out the project. Therefore,

> Residual value = Gross development value (GDV) − Total cost of development

6.3.4 Profit or account method

This method is used to estimate the rental value of premises in the case of certain properties where some element of monopoly exists. This monopoly may be legal or factual. It is a legal monopoly where some legal restraints exist to prevent competition from the property. For instance, a licence is granted to a particular premises user to carry on a specific trade that others cannot easily obtain. This method rests on the premise that the values of some properties will be related to the profits that can be made from their use. In order words, the rental value will primarily depend on the earning capacity of the property.

6.3.5 Investment method

The investment method can be used when the property produces an income. The income expected must be comparable with that which could be earned by investing the capital elsewhere. Considering alternative investment possibilities, factors such as security values, case of realisation, costs of purchase and selling, and any tax liability will influence competing proposals. The principal investors are pension funds managers, insurance companies, property companies, historic owners, local authorities and government agencies.

6.3.6 Reinstatement method
This method involves estimating the cost of rebuilding a particular property plus the value of the land on which it stands. This method works on the basic assumption of estimation of reproduction new cost as of the valuation date. Reproduction new cost, in this case, refers to the cost of creating a replica building or improvement based on current prices using the same or very similar materials. Thus, the substitute building for which the subject property is being compared should be the same in design, construction and materials. This means that the substitute or comparable building must be the same in all respects.

6.3.7 Hedonic price modelling
This is a computer-based system for valuing property based on the different variables involved. It uses the techniques of multiple regression analysis to find a formula or mathematical model that best describes the data characteristics that have been collected. The technique is normally used when the relationship between the variables is not unique. This is in the sense that one variable's value always corresponds to another's. This model can predict confidence limits to the results and, where a good model has been constructed, these should allow the value to be stated within tolerable limits. The property location is the most significant variable that affects value.

6.4. Developer's budget
Building developers searching for a good site location should consider certain factors before making a final decision. One of the factors to consider is how feasible the scheme/project is. This can be established by preparing a developer's budget. A developer's budget is the estimated cost required to take the construction project from the start to commission, including all connected costs and expenses that are accrued during the building process. In the process of the developer's budget, answers would be provided for the following questions (Ashworth, 2010).

- What would the land cost?
- What is allowed as the maximum cost of the building project?
- What would be the sale or rental value of the property?

The developer's budget includes

- gross development value
- costs of construction
- professional fees
- legal and agency fees
- cost of finance
- developer's profit.

6.4.1 Gross development value (GDV)
The GDV is the capitalisation of the net income accruing from a property. Capitalisation is the product of the net income and the year's purchase (single or dual rates). The total rental value

is estimated by comparing the proposed scheme with rents obtained from similar properties. This amount is deducted as a reasonable allowance for outgoings (maintenance, repairs, management, running cost, taxes etc.). This then provides the net income from the proposed development. The valuer will be able to advise upon these appropriate amounts. The established GDVs are, however, more prone to error than building costs. This is due to the following.

- The property market is full of doubts.
- Inadequate information; problems of information flow and fluidity; information on properties is often hoarded.
- Discrepancies in professional opinions are largely based on intuitive imagination and experience. The valuations of two independent valuers could indicate wide discrepancies.

The net income (I_{net}) is then capitalised by multiplying by an appropriate year purchase in perpetuity (YP). The YP can be obtained by dividing 100 by the interest rate (i.e. the inverse of the interest rate). So

$$\text{GDV} = I_{net} \times \text{YP}$$

This is known as the PV of ₦1 per annum (e.g. net income at ₦2000/year multiplied by YP at 8% = ₦2000 × 100/8 = ₦25 000 (development value).

One point that should be noted is that office block rents are based on net usable floor areas. Before calculating the development value, some allowances should be made for non-usable floor areas, such as circulation space.

All of the calculations shown in this chapter are in Nigerian naira (₦), unless noted otherwise

Example 1
The rental value of an office block is estimated to be ₦40/m². The total floor area is 5000 m², the non-lettable area is 25% so the usable floor area is 75%. What is the development value if the YP is 7%?

Solution 1

₦40 × 5000 m² × 75% = ₦150 000

YP at 7% = 100/7 = 14.29

GDV = ₦150 000 × 14.29 = ₦2 143 500

Example 2
Osador's and Partners, an estate surveyor and valuer, submitted the following bill to the management of Auchi Plaza. What will be the new GDV if the lettable area is 4800 m²?

Management fee per annum (p.a.) = 10% of net income

Maintenance cost p.a. = ₦180 000

Annual running cost = ₦260 000

Tenement rate/tax = ₦30 000

Rental value = ₦1500/m²

Market interest rate = 16.5%

YP = 7%

Solution 2

Hint: Deduct all outgoings from the gross income and then capitalise the net income

Outgoings

Management fee = 10% × 1500 × 4800 = 720 000

Maintenance cost p.a. = 180 000

Annual running cost = 260 000

Tenement rate/tax = 30 000

Total cost of outgoings = ₦1 190 000

$$\begin{aligned} GDV &= (I_g - C) \times YP \\ &= (1500 \times 4800 - 1190000) \times 100 / 14.29 \\ &= (7200000 - 1190000) \times 100 / 14.29 \\ &= (6010000) \times 100 / 14.29 \\ &= ₦42057382.79 \end{aligned}$$

6.4.2 Costs of construction

This is the total cost of erecting or providing a construction facility. There are several easy-to-apply methods for calculating the approximate cost of a building. However, while the methods rely upon a simple method of quantification, such as the floor area of the proposed building, the skill of selecting a correct and current rate by which to calculate cost is much more difficult. This relies on knowledge of current prices and being able to interpret these against the designer's brief and outline drawings. One of the simplest methods is the unit method of simple rate technique.

Example 3

Estimate the cost of constructing a 600 m² office building in Ewu. The cost per m² of a similar project in Ekpoma is ₦6700/m² at 2006 prices.

Solution 3

Cost of construction (Ekpoma as of 2006)

$$= ₦6700/m^2 \times 600\,m^2$$
$$= ₦4020000$$

Adjust for time and location factors:

Ewu location cost factor (assumed) = 120

Ekpoma location cost labour (assumed) = 140

2006 construction price index (assumed) = 220

2013 construction price index (assumed) = 310

The construction cost for the new building in Ewu as of 2013 as adjusted is:

= ₦4 020 000 × 120/140 (Ewu 2006)

= ₦3 445 714.29

= ₦3 445 714.29 × 310/220

= ₦4 855 324.68

= ₦8092.21/m^2

6.4.3 Professional fees

Charges for any professional services provided need to be added to the construction cost. The various professional institutions publish fee scales that can be used as a guide for assessing these costs. The fee scales are based on the construction of a lump sum and a percentage of the construction costs. The fees vary depending on the type and size of the project, and the description of the service provided. The larger the project and the more repetitive its components, the smaller the overall fee that is charged.

6.4.4 Legal and agency fees

Legal fees will be required for the purchase of the site and the preparation and agreement of leases or conveyancing documents. Property agents may also be required to let or manage the property, or be responsible for its disposal to potential owners. Their fees may be typically 2–3% of the selling price, depending on the service provided and the number of units involved.

6.4.5 Cost of finance

Before commencing with a construction project, the developer will need to have purchased a site. The developer may have chosen to use retained earnings, in which case there will be a loss of interest accruing. Alternatively, the developer may need to borrow the funds, in which case there will be an interest charge for the service. These charges will need to be added to the cost of the development. Land is often purchased at least 12 months before starting site work. The time between completion and letting or selling will vary depending on local circumstances. The interest charged is usually based on the total building cost for half a year. Although this is only an approximate figure, it is adequate for calculating the developer's budget. In the case of long contract periods, compound interest should be used. The interest rate will be based on the opportunity cost of capital, a few points higher than the base rate when finance is borrowed.

6.4.6 Developer's profit

This item allows developers a return on the project for the skills, time and risk involved. About 10–20% of the GDV should be included in the budget. The risk to be evaluated by developers and their professional advisers may include:

- rising costs
- high speculative development
- failure to lease or sell.

Example 4

A vacant plot with a frontage of 40 m and width of 30 m has planning permission for a building with certain parameters. It may cover up to two-thirds of the site, with a building plot ratio of 3. A valuer considers that the proposed building should produce a gross income of ₦110 per m^2 usable floor area. The landlord's outgoings will be ₦20 000/year, the construction cost for the building will be ₦400/m^2 and site work will cost ₦90 000. The circulation space is 15% of the gross floor area. The construction period is 2 years. Calculate the net return of the investment.

Solution 4

Floor area of building

Assume the road to the front of the site has an overall width of 10 m

Site area $= 40 \times 30 = 1200\,m^2$

Area for planning purposes includes ½ × road width $= 40 \times 35 = 1400\,m^2$

Planning area $= 1400$

Plot ratio $= 1400 \times 3 = 4200\,m^2$

Usable floor area to generate income $= 4200 \times 85\% = 3570\,m^2$

Value

Annual rental income (3570 × ₦110)	392 700
Less landlords' expense	20 000
Sum	372 700
Year of purchase in perpetuity for office, say 7%	(100/7) × 14.3
Gross development value (GDV)	5 329 610

Deduct costs

Cost of building 4200 × ₦400	1 680 000
Cost of site works	90 000
Professional fee at 10%	177 000

Finance for construction, compound at 15% p.a. (1 947 000/2 @ 15% p.a.)	313 954
Legal fees, agents fees, etc. (2 1/2% of GDV)	133 240
Developer's profit at 15% × GDV	799 442
Sum	3 193 635
Value of site plus finance	2 135 975
Cost of site finance, compound at 15% p.a. for two years = 2 135 974 × (0.3225/1.3225)	520 871
Site value (land)	1 615 104

The developer should thus pay ₦1 615 104 for the site (this would usually be rounded down to ₦1 615 000.

Construction	₦1 947 000
Finance for construction, etc.	₦313 954
Legal, etc.	₦133 240
Site	₦1 615 103
Finance for site	₦520 871
GDV minus developer's profit	₦4 530 168

A net return of ₦372 700 p.a. on ₦4 530 268 = 8.2 %

There is a need to compare the return with the prevailing returns on similar investments. This would enable the developer to determine the project's viability. From the above calculation, the developer's profit is the profit assigned for the role of a developer. Also, the building contractor's profit covers part of the construction costs for the developer.

Example 5
Below is a typical example of a budget for a residential development by a speculative developer. Calculate the minimum rental income and the minimum selling price, assuming a construction period of 9 months and ten reattached works.

(1)	Cost of land	1 500 000
(2)	Legal fees for land acquisitive	250 000
(3)	Land improvement cost	950 000
(4)	Construction cost and professional fees	29 930 000
(5)	Site layout	2 090 000
(6)	Agency fees for letting/selling	500 000
		Total 35 220 000
(7)	Cost of finance	30% p.a.

Solution 5

For items (1) (2) (3) inclusive 2 700 000 (0.30)	810 000
For items (4) (5) 32 020 000 (0.30) × 9/12 (assuming 9 months for construction)	7 204 500
For item (6), no finance is required as this is paid from the sale or leaving	8 014 500
Aggregate costs	35 220 000 + 8 014 500 = 43 234 500
Minimum annual rental income	₦43 234 500 × 0.30
Total rental income	₦12 970 350
Rental income/unit	= ₦12 970 350/10 units = ₦1 297 035
Minimum selling price	= ₦43 234 500/10 = ₦4 323 450 per unit

The value should be minimal because the budget does not include the developer's profit.

Hints

Cost of finance for construction = $(1.22)^2 - 1 \times$ cost of construction

Cost of finance for land = $(1.22)^2 - 1 \times$ cost of land

Professional fees = % of cost of the construction

Legal and advert fees = % of GDV

Developer's profit = % of GDV

Note that the interest rate could be more than 22%. Therefore, 1.22 is not a constant.

Land:

  Cost of land + cost of finance for land + legal advert fees

Construction:

  Cost of construction + professional fees + cost of finance for construction

  GDV = Land + Construction + Developer's profit
      = Land + Finance for land + Construction + Finance for construction
      + Developer's profit

Example 6

A developer wishes to erect an office with a 5000 m^2 area block along Airport Road, Benin city. The estimated construction cost is ₦4500/m^2 out of a net income of ₦2900/m^2. It is expected that the construction will take 24 months to build. The current interest rate is 22% p.a. and 80% is lettable. Determine the present market value of the site.

Solution 6

Gross development value (GDV) = Net income × YP = (2900 × 5000 × 0.80) × 100/12	₦52 727 272.73
Deduct costs	
Construction cost = 4500 × 5000	₦22 500 000
Professional fees 12% of ₦22 500 000	₦2 700 000
Cost of finance for 24 months at 22% (25 200 000(1.22)² − 1 = 25 200 000(0.4884))	₦12 307 680
Legal and agency fees 3% of GDV (₦52 727 272.73 × 0.03)	₦1 581 818.18
Profit 15% of GDV (₦52 727 272.73 × 15/100)	₦7 909 090.91
Total costs	₦46 998 589.09
Site value + site finance cost for land	₦5 728 683.64
Cost of site finance, compounded at 22% p.a. for 2 years	₦1 879 796.50
Total	₦3 848 887

Example 7

Edos Partner Limited is considering purchasing a site on which to erect 40 detached houses. The selling price of the houses is ₦65 000. The cost of the land, inclusive of legal charges, is ₦150 000. The developer requires a profit of 16% of the GDV. What is the allowable amount for the building cost?

Solution 7

GDV = Developer's gross income = 40 × 65 000	₦2 600 000
Deduct costs	
Land cost inclusive of legal charges	₦150 000
Short-term finance – assume 2.5 years at 12% compound interest rate (49 130[(1.12)^{2.5} − 1] × 150 000)	₦49 130
Legal, agents and advertising fee (3% of GDV)	₦78 000
Profit (16% of GDV)	₦416 000
Total costs	₦693 130
Sum	₦1 906 870

In summary,

GDV = ₦2 600 000

Less development costs = ₦693 130

Remainder = ₦1 906 870

Building cost

Let B = building costs. Let's assume that finance will be required for 1.5 years @ 12% and professional fees are 10%

Building cost = B
Finance = $B \times 0.12 \times 1.50$
Professional fees = $B \times 0.10$

$$₦1\,906\,870 = B + 0.10B + 0.185(1.1B)$$
$$= 1.1B + 0.185(1.1B)$$
$$= 1.1B + 0.2035B$$
$$= 1.3B$$

Therefore

$B = ₦1\,906\,870/1.3$
$B = ₦1\,466\,823.08$

Check

Land cost and finance	199 130
Legal, agent's fees	78 000
Developer's profit	416 000
Building cost	1 466 823.08
Finance	298 498.50
Fees	146 682.3
Building cost of 40 houses	₦1 466 823
Cost per house (1 466 823/40)	₦37 238/house
Assuming each house = 110 m²	
Cost per m² (₦37 237/110)	₦338.50/m²

GDV = 199 130 + 0.03GDV + 0.16GDV + 1 908 870
GDV = 2 106 000 + 0.19GDV
GDV − 0.19GDV = 2 106 000
0.81GDV = 2 106 000/0.81 = 2 600 000
Legal/agency fees = 0.02 × 2 600 000 = 78 000
Profit = 0.16 × 2 600 000 = 416 000

Sometimes, a developer can predetermine the costs and needs to know the likely selling price of the development and whether this is achievable. The questions can then be approached in the reverse order.

Land cost and finance = 199 130

Fees 3% of GDV $(X) = 0.03X$

Profit 16% of GDV $(X) = 0.16X$

Building cost, finance and fees   1 906 870

GDV X

$\text{GDV}(X)$ = 199 450 + 0.03X + 0.16X + 1 906 870

= 199 450 + 0.19X + 1 906 870

0.81X = 2 106 000

X = 2 600 000

Legal/agents fees = 0.03 × 2 600 000 = 78 0005
Developer's profit = 0.16 × 2 600 000 = 416 000

6.5. Summary

The chapter covered methods of determining the value of property and the developer's budget. This includes the factors that influence the development of construction sites – the type of development envisaged, location, shape, size, topography, aspect and access, ground conditions and site preparation difficulties, availability of utility services, planning controls, legal considerations, government assistance that might be available, the costs of developing the site and its eventual worth. Also, the chapter covered investment appraisals and identified where they could be used. This chapter concluded with the developer's budget and components with practical examples. The next chapter gives an in-depth description of the economics of sustainable construction in a developing economy.

REFERENCES

Ajayi CA (1996) Theories, techniques, and practice of development appraisal. *Presented at a National Training Workshop of the Nigerian Institution of Estate Surveyors and Valuers, Lagos, Nigeria*, pp. 1–53.

Archibong TC and Ogunba OA (2018) *An Examination of Volatility Levels of Development Variables in Uyo*. African Real Estate Society (No. afres2018_119).

Ashworth A (2010) *Cost Studies of Buildings*, 5th edn. Prentice Hall, London, UK.

Bartelink R, Appel-Meulenbroek R, van den Berg P and Gehner E (2015) Corporate real estate risks. *Journal of Corporate Real Estate* **17(4)**: 301–322, https://doi.org/10.1108/JCRE-09-2015-0020.

Fadun OS and Saka ST (2018) Risk management in the construction industry: analysis of critical success factors (CSFs) of construction projects in Nigeria. *International Journal of Development and Management Review* **13(1)**: 108–139.

Farayibi AO (2005) The impact of risk on investment decisions in Nigeria. *Research Journal of Finance and Accounting* **6(32)**: 52–59.

Gupta A and Tiwari P (2016) Investment risk scoring model for commercial properties in India. *Journal of Property Investment and Finance* **34(2)**: 156–171.

Haliza-Asat S, Nik-Wan NZ, Haron H, Jaafar M and Hassan TMRT (2017) Assessment and management of risks of housing developers: Malaysian perspectives. In *Global Conference on Business and Economics Research (GCBER), Universiti Putra Malaysia, Malaysia*.

Hallegatte S, Bangalore M and Jouanjean MA (2016) *Higher Losses and Slower Development in the Absence of Disaster Risk Management Investments*. World Bank, Washington, DC, USA.

HCL (House of Commons Library) (2019) *Permitted Development Rights*. HCL, London, UK, pp. 3-19.

Jagun ZT (2020) Risks in feasibility and viability appraisal process for property development and the investment market in Nigeria. *Journal of Property Investment & Finance* **38(3)**: 227–243, https://doi.org/10.1108/JPIF-12-2019-0151.

Khumpaisal S, Ross A and Abdulai R (2010) An examination of Thai practitioners' perceptions of risk assessment techniques in real estate development projects. *Journal of Retail and Leisure Property* **9(2)**: 151–174.

Maleta M (2012) Methods for determining the impact of the temporal trend in the valuation of land property. *Real Estate Management and Valuation* **21(2)**: 29–36.

Marshall P and Kennedy C (1993) Development valuation techniques. *Journal of Property Valuation and Investment* **11(1)**: 57–66, https://doi.org/10.1108/14635789310031423.

MHCLG (Ministry of Housing, Communities and Local Government) (2019) *Table 118: Annual Net Additional Dwellings and Components, England and the Regions, 2000-01 to 2018-19*. MHCLG, London, UK.

Newell G and Razali MN (2009) The impact of the global financial crisis on commercial property investment in Asia University of Western Sydney. *Pacific Rim Property Research Journal* **15(4)**: 430–452.

Nnamani OC (2018) *Application of Quantitative Risk Analysis in Property Development Projects in Nigeria: A Review*. European Real Estate Society, Amsterdam, the Netherlands.

Ogunbayo OT, Odebode AA, Oyedele JB and Ayodele OT (2018) The significance of real estate development process analysis to residential property investment appraisal in Abuja, Nigeria. *International Journal of Construction Management* **3599**: 1–10, https://doi.org/10.1080/15623599.2017.1423164.

Olaleye A, Aluko BT and Ajayi CA (2007) Factors influencing the choice of property portfolio diversification evaluation techniques in Nigeria. *Journal of Property Investment and Finance* **25(1)**: 23–42.

Taylor K, Edwards DJ, Lai JHK et al. (2023) Converting commercial and industrial property into rented residential accommodation: development of a decision support tool. *Facilities* **41(1/2)**: 1–29, https://doi.org/10.1108/F-01-2022-0006.

Wiegelmann TW (2012) *Risk Management in the Real Estate Development Industry*. PhD thesis, Institute of Sustainable Development & Architecture, Bond University, Robina, Australia.

Principles of Basic Construction Economics in the 21st Century

Andrew Ebekozien and Clinton Aigbavboa
ISBN 978-1-83549-841-5
https://doi.org/10.1108/978-1-83549-838-520241007
Emerald Publishing Limited: All rights reserved

Chapter 7
Economics of sustainable construction

7.1. Introduction to sustainable construction

Physical infrastructure provision in large-scale development is of paramount importance, especially in developing countries' new layouts. The construction of infrastructure comes at a cost to the natural environment, especially when conventional construction methods are used. Traditional construction methods are unsustainable (Tunji-Olayeni et al., 2020) as they deplete limited natural resources such as timber and freshwater. For example, freshwater is in high demand during the construction and occupancy phases. Adzawla et al. (2019) argued that the dependence on fossil fuel as a cheap energy source in residential buildings remains a concern because it is one of the major sources of greenhouse gases. This is complicated by the continuous arbitrary felling of trees in forests for construction and other purposes. This threatens the environment and contributes significantly to global warming (Ebekozien et al., 2022a; Whitehead, 2011). The provision of facilities should be done in a way that is ecologically and socially responsible in order to mitigate the 'cobra effect', which is a term used to describe attempted measures/solutions that worsen a present issue. To achieve this task, sustainable interventions (e.g. green building, green construction, sustainable architecture, alternative building technologies and practices) are being encouraged and embraced. The terms sustainable architecture, eco-build, green build, green construction, sustainable materials, ecological buildings and green buildings are not new terms in the concept of sustainable development (Kibert, 2007). The construction industry is being given special attention because it is key to achieving the United Nations Sustainable Development Goals. The world is facing increasing global warming, deforestation and the depletion of natural resources (Ebekozien et al., 2023a; Isang, 2023). Increased awareness by the World Commission on Environment and Development has awakened nations to the necessity of preserving the environment. In many developing countries, including Nigeria, understanding and awareness of sustainable construction were missing in the 20th century, but studies by, for example, Dania et al. (2013) and Isang (2016) have significantly and profoundly contributed to Nigeria's sustainable development. These studies are among the top scholarly literature emphasising Nigeria's need for sustainable construction.

As cited by Bourdeau (1999: p. 41), Charles Kibert was the first scholar to define sustainable construction as

> the creation and responsible management of a healthy built environment based on resource efficient and ecological principles

Other acknowledged definitions include (Lanting, 1998: p. 6)

> a way of building which aims at reducing (negative) health and environmental impacts caused by the construction process or by buildings or the built environment

Du Plessis (2002: p. 8) defined sustainable construction as

> a holistic process aiming to restore and maintain harmony between the natural and the built environments and create settlements that affirm human dignity and encourage economic equity

These definitions imply that sustainable construction is all about the construction, operation and maintenance of structures that meet clients' and end-users' demands over their lifespan, have reduced negative environmental impacts and promote cultural, social and economic progress. However, none of these definitions is exclusively satisfactory.

The overall concept of sustainable construction ensures that the industry attains sustainability (Hill and Bowen, 1997; Willar et al., 2021). There are several sustainable construction practices because the concept is still evolving. Material reuse, energy efficiency, waste management and compliance with health and safety regulations were identified by Tunji-Olayeni et al. (2020) as common sustainable practices in Nigeria. Omopariola et al. (2022) identified material waste, non-management of workers' health and safety, unsustainable technologies, negative externalities and excess energy consumption as unsustainable practices in Nigeria. Sustainable practices consider topics like waste minimisation, productivity, efficiency and safety (Abd Jamil and Fathi, 2016) in concurrence with a project's economic, social and ecological issues (Shurrab et al., 2019). A refusal to change from these unsustainable practices could lead to environmental pollution, which can trigger respiratory illnesses such as lung cancer, bronchitis, asthma and other allergies, which can lead to death (Jiang et al., 2016; Kim et al., 2018).

7.2. Benefits of sustainable construction to the industry

The benefits of sustainable construction supersede the initial high-cost implications and mitigate the effect of project actions on the environment and the economy. This necessitates the demand for green policies in Nigeria. Progress in this direction has been slow because of the dependence on the government for policy directions. Although the industry is slow to respond to novelties in sustainable construction (Ebekozien et al., 2022a, 2022b; Ercan, 2019), it has started to employ sustainable construction to mitigate the economic, social and environmental problems that can arise from construction activities. There are some green building policies in Nigeria, including the National Adaptation and Plan, the Nigeria Building Code, the National Building Efficiency Code and the Green Building Council of Nigeria (GBCN) (Abisuga and Okuntade, 2020). As of 2022, the GBCN attained World Green Building Council membership status (GBCN, 2024). The benefits identified by Nduka and Sotunbo (2014) and Ebekozien et al. (2022a, 2023a) include improving productivity, decreasing environmental damage, preventing/mitigating global warming, conserving natural resources and active recycling. Isang (2016, 2023) noted the economic, social and environmental benefits to be enjoyed by stakeholders, including clients, end-users, contractors and construction workers. These benefits include increased building value, lower construction and operating costs, higher productivity/

performance and improved health. Sustainable construction projects can also enhance natural debasement, reduce waste, secure residents' health and well-being, optimise energy and water usage, create low maintenance and lower operating risk (Abisuga and Okuntade, 2020).

Among the perceived benefits of sustainable construction in Nigeria are increased jobs/opportunities, reduction of future maintenance costs, greater productivity and end-users' satisfaction, improvements in internal air quality and enhancements in inhabitants' comfort and health (Alohan and Oyetunji, 2021). Similarly, Okoye *et al.* (2021) noted that sustainable construction improves indoor environmental quality, reduces operational and maintenance costs, mitigates the impacts of climate change and global warming, minimises construction waste and increases profitability and competitive advantages. Sustainable construction practices increase building functionality, durability and energy efficiency, and reduce environmental impacts (Ebekozien *et al.*, 2022a). Sustainable construction practices should be implemented to improve the environment and should be integrated with economic and social issues in order to create better and more efficient lives (Abd Jamil and Fathi, 2016). Regarding construction projects in Nigeria that have expressed the socio-economic and environmental benefits of sustainable construction, Isang (2023) identified the Wing and Alliance building in Lagos, the Heritage building, the Nestoil Tower, Abacus One Estate in Abuja and Primetech's head office.

7.3. Principles of sustainable construction

This section discusses the three main aspects of sustainable construction practice (economic, social and environmental sustainability).

7.3.1 Environmental sustainability

Environmental sustainability is one of the three pillars of sustainable development. It intends to guide society to meet its requests without weakening biological diversity or preventing the ecosystem from renewal (Morelli, 2011). It should preserve the qualities of the physical environment (Sutton, 2004) and focus on mitigating tasks that influence environmental quality in the long-term (Pero *et al.*, 2017). Practitioners should thus be encouraged to use sustainable construction materials and project methods. Unsustainable mechanisms (e.g. the use of landfills for construction waste, full dependence on fossil fuel, indiscriminate tree felling and unregulated extraction of limestone) can trigger environmental problems for humans and the planet as a whole.

7.3.2 Social sustainability

Social sustainability is another key aspect of sustainable development. Different scholars (Almahmoud and Doloi, 2018; Davoodi *et al.*, 2014; Littig and Grießler, 2005; Polese and Stren, 1999; Sachs, 1999; Woodcraft *et al.*, 2011) have conceptualised it differently. According to Polese and Stren (1999), social sustainability is meeting people's cultural and socially diverse needs while safeguarding their compatible cohabitation in a place. Sachs (1999) described it as regarding equity and fairness in allocating and distributing social goods for creating decent and liveable societies. Littig and Grießler (2005) emphasised that social equity and integration in assessing basic amenities are critical variables in defining social sustainability. This was corroborated by Woodcraft *et al.* (2011), who noted that communities' needs are germane in achieving social sustainability. Regarding integrating social sustainability into design, Davoodi *et al.* (2014) opined that, to fulfil the criteria, participatory design, architectural identity, hierarchy, flexibility, social security and social interaction should be considered. Social sustainability can be integrated into construction projects by meeting the demands of neighbourhood communities, end-users and industry actors (Almahmoud and Doloi, 2018).

7.3.3 Economic sustainability

The third pillar is economic sustainability. Many view this as the affordability of environmental and social outcomes (Tunji-Olayeni *et al.*, 2020). The high cost of construction is leading contractors to consider ways to mitigate costs. Conventional construction methods focus mainly on the initial construction cost, without thought of the other costs incurred throughout a project's life cycle. Tunji-Olayeni *et al.* (2020) affirmed that the conventional approach promotes higher operational and maintenance costs and, most times, increases greenhouse gas emissions. It is thus necessary to explore mechanisms to appraise alternative construction practices concerning the benefits and costs over a defined duration (Dwaikata and Alib, 2018). This is called life-cycle costing (refer to Chapter 4) and is a reliable economic tool for evaluating criteria in the energy and construction sectors (Naves *et al.*, 2019). Economic sustainability also concerns stakeholders' profitability and project viability, which includes affordability for users without jeopardising quality.

7.4. Encumbrances to economic sustainable construction

Despite the benefits of sustainable construction to the industry, the application of sustainable construction technologies is still low. Sustainable construction faces economic issues at different phases. Bon and Hutchinson (2000) and Ebekozien *et al.* (2023a) acknowledged that implementation issues are high in less developed countries (LDCs) and newly industrialised countries (NICs). This section discusses the barriers to sustainable construction at the economic analysis levels (macro, meso and micro) identified by Bon and Hutchinson (2000). Encumbrances facing each economic analysis level will be discussed in turn, with the intention of increasing stakeholders' attention to broad classes of issues begging for solutions regarding fruitful implementation of the sustainable construction agenda.

Economic aspects or features are key positive catalysts for achieving sustainable development goals or sustainable construction. It has been argued that governments' economic measures are more effective than restrictions and prohibition measures in achieving sustainable construction projects (Ebekozien *et al.*, 2022a, 2022b, 2023a, 2023b). Most economic measures should be initiated as government interventions or social interests. Hence, the role of governments should be guided by a sustainable and integrated framework to enhance continuity.

7.4.1 Macroeconomic encumbrances

Sustainability has become a global issue across many sectors, including the construction industry. From a global perspective, one issue with sustainable construction is the global supply of construction action (Bon and Crosthwaite, 2000). Bon and Hutchinson (2000) argued that countries where construction activities are growing are not suitable for the implementation of sustainable construction. This is debatable. Bon and Hutchinson (2000) opined that the industry might only be a strong, sustainable development contributor if critical actions are taken to revise the trend. Rees (1999) asserted that power booms and materials consumption in LDCs and NICs are offset by the significant 'dematerialisation' of construction action in developed industrialised countries (DICs). Complementing this requires sustainable development projects in DICs to advance quicker than anticipated in order to accommodate the slower economic growth in NICs and LDCs.

7.4.2 Mesoeconomic encumbrances

The construction industry is composed of typical assembly industries, so the industry buys from other industries on the mesoeconomic level. Changing its output attributes is thus difficult. It is an assembler industry. Bon and Hutchinson (2000) opined that, besides choosing green materials and assembling methods, it is germane to establish whether the materials and methods are actually green. This is missing in many construction projects in LDCs and NICs. Experience shows that non-green inputs may only be identifiable within the supply chain once they arrive at the construction site. Also, no established mechanism ensures all effects will be covered at a given aggregation level. Hence, measures to promote sustainable development should be encouraged to mitigate the challenge.

7.4.3 Microeconomic encumbrances

Business life correlates with building life and other interconnects. As business horizons shrink, building lives shrink (Bon, 1989). People demand buildings because of the services the building will offer – people do not want buildings but their services. These services are connected with economic progressions that create various goods and services. In sustainable construction, there is more growth in the share of electronic, electrical and mechanical equipment in building investment than in blocks/bricks. For sustainable resource management, there is an uncertainty challenge in product markets, leading to a tendency for the life cycle of production equipment and physical plant to converge. This may have contributed to why constructed facilities are erected within shorter time horizons because of the uncertain economic environment and increased digitalisation of installation/construction. This is a call for all concerned because a sustainable development/construction goal should be a long-term outlook.

7.5. Encumbrances of sustainable construction in developing countries

There are several issues confronting the implementation of sustainable construction, especially in NICs and LDCs (Bon and Hutchinson, 2000; Du Plessis, 2007; Ebekozien *et al.*, 2023a; Kibert, 2007; Rees, 1999). This section focuses on developing countries or LDCs. The systematic issues confronting LDCs make it complicated to implement sustainable construction projects. Du Plessis (2007) and Ebekozien *et al.* (2023b) identified institutional incapacity, weak governance, environmental degradation, an uncertain economic environment, lax skills levels, social inequity, abject poverty and fast urbanisation rates as the issues facing LDCs. These issues create an anti-enabling environment for the implementation of sustainable construction projects. These issues thus threaten sustainable development's main pillars (social, economic and environmental), but also extend to other extended pillars (institutional, political and technical) (Hill and Bowen, 1997).

In many LDCs, including Nigeria, it has been challenging to balance the relationship between creating jobs from construction projects and protecting the environment or between a renewable energy project and its social and environmental impact and bringing people near nature. The three pillars model attempts to address these issues, but there are challenges (Barbier, 1987). A multi-faceted approach and holistic thinking are needed. The absence of a holistic strategy to contribute to human, economic and physical development is a challenge. Abolore (2012), Dania et al. (2013) and Ebekozien et al. (2023a) found that knowledge and understanding of sustainable construction is low in Nigeria. Other barriers to sustainable construction in Nigeria have been found to be (Akinshipe et al., 2019; Ebekozien et al., 2022a, 2023a; Udo and Udo, 2020)

- resistance to cultural change
- ignorance of the benefits of life-cycle costing
- poor coverage of sustainability in higher educational institutions
- poor demand for sustainable construction products
- inadequate techniques to promote green building practices
- inadequate knowledge and expertise.

In addition, Oribuyaku (2015) and Ebekozien et al. (2022a) identified the absence of a sustainable building code for practitioners in the industry, while Babalola (2020) and Ebekozien et al. (2022a) suggested an integrated green policy framework for sustainable construction implementation practices is needed. Ebekozien et al. (2022a, 2023a) and Isang (2023) noted lax government participation at all levels (federal, state and local) and scant enforcement of policies regarding sustainable construction and green building practices in Nigeria. An absence of public agencies to enforce green practices on building projects and the high cost of sustainable constructions have also been found to be major barriers (Ebekozien et al., 2022a; Onuoha et al., 2017; Osuizugbo et al., 2020). Ebekozien et al. (2022a) identified the absence of a standard framework for green certification of buildings as a major challenge to sustainable construction implementation in Nigeria. As a result, they developed a model for promoting green certification of buildings. Eight sub-themes were clustered to increase green certification of buildings in developing countries, using Nigeria as a case study.

The governments of developing countries, including Nigeria, should do more to advance sustainable construction. The Nigerian government has made some progress in the past few years, but there is space for improvement. Babalola (2020), Ebekozien et al. (2022a, 2023a), and Isang (2023) asserted that the Nigerian federal government is advancing sustainable construction practices and green building construction. Strategies to motivate stakeholders to embrace the practices are being improved (Omopariola et al., 2022), which has led to some green building policies, such as the National Adaptation Plan, the Nigeria Building Code, the National Building Efficiency Code and the GBCN.

7.6. The way forward

As part of the way forward, the following are recommended.

- The importance of making a radical shift in construction practices and methods – from pre- to post-construction – cannot be over-emphasised. Several studies have shown that the construction sector is a major contributor to environmental damage and global climate change. Governments need to revamp policies that are pro-change and tailored towards improving sustainable construction projects. In addition, civic awareness among stakeholders to build sustainably – now and in the future – should be all-inclusive. Civil society should reawaken, organise protests and express their displeasure for unsustainable construction firms or materials producers.

- Governments need to take the lead in governance through standards, market-oriented policies and programmes, and legal and regulatory practices to positively influence and attract sustainable construction investors and push construction companies towards sustainable construction practices. Such actions should be complemented with incentives like tax reliefs, subsidies and soft loans to clients and contractors embracing sustainable construction (Ebekozien *et al.*, 2023a; Tunji-Olayeni *et al.*, 2020). Market-oriented action is key to attracting foreign investors. Thus, government interventions in collaboration with international organisations like the International Council for Research and Innovation in Building and Construction (CIB) are critical at every phase to produce the necessary outcomes. The CIB is one of the largest construction-related international organisations and has shown concern with an active role in sustainable construction matters (Bon and Hutchinson, 2000). Government regulation should be given the needed attention to improve achieving sustainable construction (Oladokun *et al.*, 2017).

- The construction industry's main stakeholders, especially contractors, should be willing to do the responsible thing and acknowledge that it is no longer about selection. The change should be seen from a moral imperative and partnerships to protect biodiversity and conserve natural resources. A framework to articulate this moral imperative is germane. The model should address issues surrounding benchmark information, network building and partnerships, access to funding, promoting awareness and capacity building. A workable framework involves collaboration and consultation/dialogue among key stakeholders, including research institutions/centres and civil society at international, regional and national levels.

- Understanding the multi-faceted connections between processes and methods that underpin the industry's activities, including construction and maintenance, is germane to improving the achievement of sustainable construction projects. This would also strengthen the implementation of developed regulatory mechanisms using institutions and partnership collaborations at all levels, with an emphasis on the roles and responsibilities of stakeholders. A policy framework tailored towards a new sustainable code for the industry's professionals is important (Oribuyaku, 2015). It is vital for construction stakeholders to have an in-depth knowledge of sustainable construction drivers and policies in order to improve sustainable practices (Darko *et al.*, 2017; Oke *et al.*, 2018).

- Developing accessible and sustainable funding streams and methods, and supporting them with tools and mechanisms to monitor and evaluate organisations' performance, are key enablers that could create a supportive environment for sustainable construction. Adjusting construction funding to short and medium terms conflicts with sustainable construction principles, which emphasise long-term benefits.

- The importance of continuous training (upskilling and reskilling) of practitioners (Oladokun *et al.*, 2017) and research into alternative building technologies and green building materials in the context of the greening revolution and innovation in collaboration with stakeholders (industry and academic) cannot be over-emphasised. The National Building and Road Research Institute in Nigeria should be restarted and funded to organise workshops and practical classes for stakeholders. This would go a long way in improving sustainability education in higher educational institutions (Toriola-Coker et al., 2021) and mitigating inadequate knowledge and expertise (Daniel et al., 2018; Udo and Udo, 2020). Also, more awareness of the implications or dangers of non-sustainable practices in the industry should be encouraged.

7.7. Summary

This chapter covered the role of building economics in sustainable construction. It covered the concept, characteristics, evolution and sustainable construction elements. This includes cost planning, cost monitoring, cost control of sustainable construction and the cost of sustainability (environmental, social and economic costs). Green construction is confronted with economic issues at different phases. The chapter identified the challenges associated with the three main levels – macroeconomic, mesoeconomic and microeconomic. The three main aspects of sustainable construction were discussed. It was concluded that market-oriented measures and strategies would improve the implementation of sustainable construction practices. The recommended strategies and measures highlighted in Section 7.6 include a national action plan that is all-inclusive of key stakeholders' collaboration and communication, a legal framework to enact sustainable construction or a 'Green Building Act', sustainable construction incentives to motivate stakeholders, increased awareness of sustainable construction with emphasis on the benefits, the encouragement of research regarding locally based rating tools and the establishment of a well-funded government agency to manage an institutional framework for sustainability and continuity. The next chapter gives an in-depth description of the economics of smart construction in a developing economy.

REFERENCES

Abd Jamil AH and Fathi MS (2016) The integration of lean construction and sustainable construction: a stakeholder perspective in analysing sustainable lean construction strategies in Malaysia *Procedia Computer Science* **100**: 634–643.

Abisuga AO and Okuntade TF (2020) The current state of green building development in Nigerian construction industry: policy and implications. In *Green Building in Developing Countries. Green Energy and Technology* (Gou Z (ed.)). Springer, Cham, Switzerland, pp. 129–146.

Abolore AA (2012) Comparative study of environmental sustainability in building construction in Nigeria and Malaysia. *Journal of Emerging Trends in Economics and Management Sciences* **3(6)**: 951–961, https://doi.org/10.10520/EJC130248.

Adzawla W, Sawaneh M and Yusuf AM (2019) Greenhouse gases emission and economic growth nexus of sub Saharan Africa. *Scientific African* **3**: e00065.

Akinshipe O, Oluleye IB and Aigbavboa C (2019) Adopting sustainable construction in Nigeria: major constraints. *IOP Conference Series: Materials Science and Engineering* **640**: 1–5, https://doi.org/10.1088/1757-899X/640/1/012020.

Almahmoud E and Doloi HK (2018) Assessment of social sustainability in construction projects using social network analysis. *Journal of International Business Research and Marketing* **3(6)**: 35–46.

Alohan EO and Oyetunji AK (2021) Hindrance and benefits to green building implementation: evidence from Benin City, Nigeria. *Real Estate Management and Valuation* **29(3)**: 65–76, https://doi.org/10.2478/remav-2021-0022.

Babalola AA (2020) *A Policy Framework for the Implementation of Sustainable Construction Practice in Nigeria*. PhD thesis, University of KwaZulu-Natal, Durban, South Africa. See https://researchspace.ukzn.ac.za/items/3d4b99af-5c65-4dfb-9fc6-f24215a19f6a (accessed 15/04/2024).

Barbier EB (1987) The concept of sustainable economic development. *Environmental Conservation* **14(2)**: 101–110.

Bon R (1989) *Building as an Economic Process: An Introduction to Building Economics*. Prentice-Hall, Englewood Cliffs, NJ, USA.

Bon R and Crosthwaite D (2000) *The Future of International Construction*. Thomas Telford Publishing, London, UK.

Bon R and Hutchinson K (2000) Sustainable construction: some economic challenges. *Building Research & Information* **28(5-6)**: 310–314, https://doi.org/10.1080/096132100418465

Bourdeau L (1999) *Agenda 21 on Sustainable Construction*. CIB, Rotterdam, the Netherlands, CIB Report Publication 237.

Dania AA, Larsen GD and Yao R (2013) *Sustainable Construction in Nigeria: Understanding Firm Level Perspectives*. See https://www.academia.edu/4022647 (accessed 15/04/2024).

Daniel EI, Oshineye O and Oshodi O (2018) Barriers to sustainable construction practice in Nigeria. In *Proceedings of the 34th Annual ARCOM Conference*, Belfast, UK (Gorse C and Neilson CJ (eds)). ARCOM, Manchester, UK, pp. 149–158.

Darko A, Zhang C and Chan APC (2017) Drivers for green building: a review of empirical studies. *Habitat International* **60**: 34–49.

Davoodi S, Fallah H and Aliabadi M (2014) Determination of affective criterion on social sustainability in architectural design. In *Proceedings of the 8th SASTech 2014 Symposium on Advances in Science and Technology-Commission-IV, Mashhad, Iran*.

Du Plessis C (2002) *Agenda 21 for Sustainable Construction in Developing Countries*. CSIR, CIB and UNEP-IETC, Pretoria, South Africa, CSIR Report BOU/E0204.

Du Plessis C (2007) A strategic framework for sustainable construction in developing countries. *Construction Management and Economics* **25(1)**: 67–76, https://doi.org/10.1080/01446190600601313.

Dwaikata LN and Alib KN (2018) Green buildings life cycle cost analysis and life cycle budget development: practical applications. *Journal of Building Engineering* **18**: 303–311.

Ebekozien A, Ayo-Odifiri OS, Nwaole CNA, Ibeabuchi LA and Uwadia EF (2022a) Barriers in Nigeria's public hospital green buildings implementation initiatives. *Journal of Facilities Management* **20(4)**: 586–605, https://doi.org/10.1108/JEM-01-2021-0009.

Ebekozien A, Aigbavboa C, Thwala WD *et al.* (2022b) A systematic review of green building practices implementation in Africa. *Journal of Facilities Management* **22(1)**: 91–107, https://doi.org/10.1108/JFM-09-2021-0096.

Ebekozien A, Aigbavboa C and Samsurijan MS (2023a) Appraising alternative building technologies adoption in low-cost housing provision to achieving Sustainable Development Goal 11. *Engineering, Construction and Architectural Management* **31(13)**: 41–58, https://doi.org/10.1108/ECAM-06-2023-0538.

Ebekozien A, Aigbavboa C, Samsurijan MS, Adjekophori B and Nwaole A (2023b) Leakages in affordable housing delivery: threat to achieving Sustainable Development Goal 11. *Engineering, Construction and Architectural Management*. https://doi.org/10.1108/ECAM-08-2022-0758.

Ercan T (2019) *Building the Link Between Technological Capacity Strategies and Innovation in Construction.* See https://doi.org/10.5772/intechopen.88238 (accessed 15/04/2024).

GBCN (Green Building Council Nigeria) (2024) See https://gbcn.org.ng (accessed 15/04/2024).

Hill RC and Bowen PA (1997) Sustainable construction: principles and a framework for attainment. *Construction Management and Economics* **15(3)**: 223–239.

Isang WI (2016) *Appraisal of the Implementation of Sustainability Practices During Construction Phase of Building Projects in Akwa Ibom State.* MSc dissertation, University of Uyo, Uyo, Nigeria. See https://www.academia.edu/43407030 (accessed 15/04/2024).

Isang WI (2023) A historical review of sustainable construction in Nigeria: a decade of development and progression. *Frontiers in Engineering and Built Environment* **3(3)**: 206–218, https://doi.org/10.1108/FEBE-02-2023-0010

Jiang XQ, Mei XD and Feng D (2016) Air pollution and chronic airway diseases: what should people know and do? *Journal of Thoracic Disease* **8(1)**: E31–E40, https://doi.org/10.3978/j.issn.2072-1439.2015.11.50.

Kibert JC (2007) The next generation of sustainable construction. *Building Research & Information* **35(6)**: 595–601, https://doi.org/10.1080/09613210701467040.

Kim D, Chen Z, Zhou LF and Huang SX (2018) Air pollutants and early origins of respiratory diseases. *Chronic Diseases and Translational Medicine* **4(2)**: 75–94 https://doi.org/10.1016/j.cdtm.2018.03.003.

Lanting R (1998) *Sustainable Construction in the Netherlands, Report 9 in CIB Sustainable Development and the Future of Construction. A Comparison of Visions from Various Countries.* CIB, Rotterdam, the Netherlands, CIB Report Publication 225.

Littig B and Grießler E (2005) Social sustainability: a catchword between political pragmatism and social theory. *International Journal of Sustainable Development* **8(2)**: 65–79.

Morelli J (2011) Environmental sustainability: a definition for environmental professionals. *Journal of Environmental Sustainability* **1(1)**: 2, https://doi.org/10.14448/jes.01.0002.

Naves AX, Barreneche C, Fernandez AI et al. (2019) Life cycle costing as a bottom line for the life cycle sustainability assessment in the solar energy sector: a review. *Solar Energy* **192**: 238–262.

Nduka DO and Sotunbo AS (2014) Stakeholders perception on the awareness of green building rating systems and accruable benefits in construction projects in Nigeria. *Journal of Sustainable Development in Africa* **16(7)**: 118–130.

Oke A, Aghimen D, Aigbavboa C and Musenga C (2018) Drivers of sustainable construction practices in the Zambian construction industry. *Energy Procedia* **158**: 3246–3252.

Okoye PU, Odesola IA and Okolie KC (2021) Optimising the awareness of benefits of sustainable construction practices in Nigeria. *Journal of Real Estate Economics and Construction Management* **9**: 62–77, https://doi.org/10.2478/bjreecm-2021-0006.

Oladokun MG, Aigbavboa CO and Isang IW (2017) Evaluating the measures of improving the implementation of sustainability practices on building projects in Akwa Ibom State, Nigeria. *In Proceedings of Environmental Design and Management International Conference, Obafemi Awolowo University, Nigeria.* See https://www.academia.edu/38808908 (accessed 15/04/2024).

Omopariola ED, Olanrewaju OI, Albert I, Oke AE and Ibiyemi SB (2022) Sustainable construction in the Nigerian construction industry: unsustainable practices, barriers and strategies. *Journal of Engineering, Design and Technology*, https://doi.org/10.1108/JEDT-11-2021-0639.

Onuoha IJ, Kamarudin N, Aliagha GU et al. (2017) Developing policies and programmes for green buildings: what can Nigeria learn from Malaysia's experience? *International Journal of Real Estate Studies* **11(2)**: 50–58.

Oribuyaku D (2015) *Code for a Sustainable Built Environment in Nigeria: A Proposed High-level Vision of a Policy Framework*. University Library of Munich, Munich, Germany, MPRA Paper 66197.

Osuizugbo IC, Oyeyipo O, Lahanmi A, Morakinyo A and Olaniyi O (2020) Barriers to the adoption of sustainable construction. *European Journal of Sustainable Development* **9(2)**: 150–162, https://doi.org/10.14207/ejsd.2020.v9n2p150.

Pero M, Moretto A, Bottani E and Bigliardi B (2017) Environmental collaboration for sustainability in the construction industry: an exploratory study in Italy. *Sustainability* **9(125)**: 1–25, https://doi.org/10.3390/su9010125.

Polese M and Stren R (1999) *The Social Sustainability of Cities: Diversity and the Management of Change*. University of Toronto Press, Toronto, Canada.

Rees WE (1999) The built environment and the ecosphere: a global perspective. *Building Research and Information* **27(4–5)**: 206–220.

Sachs I (1999) Social sustainability and whole development: exploring the dimensions of sustainable development. In *Sustainability and the Social Sciences. A Cross-Disciplinary Approach to Integrating Environmental Considerations Into Theoretical Reorientation* (Becker E and Jahn T (eds)). ZED Books, London, UK.

Shurrab J, Hussain M and Khan M (2019) Green and sustainable practices in the construction industry: a confirmatory factor analysis approach. *Engineering Construction and Architectural Management* **26(6)**: 1063–1086.

Sutton P (2004) *A Perspective on Environmental Sustainability?* See http://www.green-innovations.asn.au/A-Perspective-on-Environmental-Sustainability (accessed 03/10/2023).

Toriola-Coker LO, Alaka H, Bello WA et al. (2021) Sustainability barriers in Nigeria construction practice. *IOP Conference Series: Materials Science and Engineering* **1036**: 012023, https://doi.org/10.1088/1757-899X/1036/1/012023.

Tunji-Olayeni PF, Kajimo-Shakantu K and Osunrayi E (2020) Practitioners' experiences with the drivers and practices for implementing sustainable construction in Nigeria: a qualitative assessment. *Smart and Sustainable Built Environment* **9(4)**: 443–465, https://doi.org/10.1108/SASBE-11-2019-0146.

Udo IE and Udo NE (2020) Major constraints affecting the sustainability of construction in Akwa Ibom State. *Journal of Environmental Design and Construction Management* **19(4)**: 267–274.

Whitehead D (2011) Forests as carbon sinks—benefits and consequences. *Tree Physiology* **31(9)**: 893–902.

Willar D, Waney YVE, Pangemanan GDD and Mait GER (2021) Sustainable construction practices in the execution of infrastructure projects: the extent of implementation. *Smart and Sustainable Built Environment* **10(1)**: 106–124, https://doi.org/10.1108/SASBE-07-2019-0086.

Woodcraft S, Hackett T and Caistor-Arendar L (2011) *Design for Social Sustainability: A Framework for Creating Thriving New Communities*. Future Communities, The Young Foundation, London, UK.

Principles of Basic Construction Economics in the 21st Century

Andrew Ebekozien and Clinton Aigbavboa
ISBN 978-1-83549-841-5
https://doi.org/10.1108/978-1-83549-838-520241008
Emerald Publishing Limited: All rights reserved

Chapter 8
Economics of smart construction

8.1. Introduction and foundations of smart construction

Unfortunately, the construction industry is one of the top industries regarding accidents. In China, the construction industry's recorded accidents have topped trade, industrial and mining accidents for about a decade. In China, in mid-2018, 1732 accidents and 1752 deaths had occurred in workplaces (SCOSCC, 2018). In 2017, the USA's construction sector recorded 971 deaths (the highest fatality), as reported by the US Department of Labor (Jiang et al., 2021). The UK is not exempt from construction accidents (Jiang et al., 2021). Malik et al. (2019) argued that the uniqueness of constructions and the outdoor environment of major activities contribute to the strong risk and randomness of accidents. Dester and Blockley (1995) and Jiang et al. (2021) opined that low automation levels at construction sites and the unsafe behaviour of workers could lead to site accidents. To mitigate construction accidents, Industry 4.0 (increasing use of information and communication technology (ICT) in the construction industry) or smart technologies through a pervasive computing paradigm should be embraced on construction sites. Both concepts involve increasing automation and digitalisation of the construction environment and creating a digital value chain for effective communication. Governments should lead research initiatives and funding programmes to achieve this goal – and this is the case in many countries. In an era of fast innovation (mobile computing enables connectivity anywhere, anytime) and developing smart technologies (with embedded intelligence and a pervasive computing paradigm), built environment activities cannot be left behind (Edirisinghe, 2019). Bowden et al. (2006) and Oesterreich and Teuteberg (2016) argued that technological trends are setting a new direction for the construction industry. Carbonari et al. (2011) asserted that real-time embedded applications characterise these trends. Smart construction sites can accomplish high-quality construction and a safe working environment because the mechanism allows for the identification, warning and control of different objects, allowing integrated security analysis systems and automated data collection (Zhou et al., 2018).

The early 21st century embraced a rapid shift from the mobile computing paradigm to the pervasive computing paradigm. Weiser et al. (1999) defined pervasive computing as 'the physical world that is richly and invisibly interwoven with sensors, actuators, displays, and computational elements, embedded seamlessly in the everyday objects of our lives, and connected through a continuous network.' Edirisinghe (2019) described it as omnipresent computing envisions linking beyond conventional devices. Chips are embedded in the connectivity notion of a device's vision, but the human body, coffee mugs, tools and clothing can serve as devices to embed chips. Gubbi et al. (2013) opined that pervasive applications combine technologies like artificial intelligence (AI), internet capability, voice recognition and wireless computing to enable devices to be modified to be 'smart'. The fundamental aspect of a smart system is 'context awareness'. According to Dey (2001), 'A system is context-aware if it uses context to provide relevant information and services to the user, where relevance depends on the user's task'. In this section, context implies details that can describe an entire scenario.

Smart construction is a concept that integrates digitalisation processes to improve productivity and output. Mihindu and Arayici (2008) asserted that smart construction is an initiative to improve engineering/construction outputs. Sutrisna *et al.* (2015) identified ICT and Internet of Things (IoT) applications as germane to smart construction sites. In addition, cloud computing and big data are enabling factors for the adoption of smart construction (Oesterreich and Teuteberg, 2016). Real-time sensor information is an example of ICT that can help construction workers realise the risk of danger. Ultra-wideband (UWB), global positioning systems (GPS), sonar and radar are examples of sensing technologies used for proximity warning systems and risk monitoring equipment (Jiang *et al.*, 2021; Kim *et al.*, 2015; Lee *et al.*, 2012). These technologies leverage the life-cycle systems of building information modelling (BIM). Meng *et al.* (2020) found 24 BIM applications in a building's life cycle. Despite the benefits of BIM (e.g. digital management of the life cycle), its application comes with challenges and risks. Chen *et al.* (2022) clustered 26 technologies into design and construction automation, communication, visualisation, analysis and data acquisition. Smart construction can improve construction/engineering project delivery, including maintenance activities (Fang *et al.*, 2018; Li, 2017; Mills, 2016). Zhou *et al.* (2018) and Woodhead *et al.* (2018) affirmed that smart construction sites are IoT ecosystems, with the concept improving construction project management through engineering data accumulation.

Several studies have been conducted regarding smart construction, which is still evolving. Niu *et al.* (2017) developed a proactive data management system for construction site facilities to improve equipment and operations efficiency. A smart construction site for road maintenance safety using robots and IoT to reduce human injuries was investigated by Li (2017). Kochovski and Vlado (2018) planned a management strategy to enhance site facilities, achieved through a computing model supporting smart construction site applications. Zhou *et al.* (2018) proposed a smart construction framework driven by ICT, using the Hong Kong–Zhuhai–Macao Bridge as a case study.

Despite the benefits for construction industry stakeholders, current applications of smart construction are low, especially in developing countries. Oesterreich and Teuteberg (2016) identified the major features required for smart construction implementation. These include networked manufacturing systems and vertical integration via the value web. Digital integration and processes of IT systems, including value chains using cyber–physical systems, are germane. Oesterreich and Teuteberg (2016) clustered key technologies connected with Industry 4.0 into smart factories, simulation and modelling, and digitalisation and virtualisation.

The technologies used in smart factories include

- radio frequency identification (RFID)
- IoT and Internet of Senses (IoS)

- automation
- modularisation/prefabrication
- additive manufacturing
- product life-cycle management
- robotics
- human–computer interaction
- cyber–physical systems/embedded systems.

The technologies enlisted in simulation and modelling include

- augmented/virtual/mixed reality
- BIM
- simulation tools and simulation models.

The technologies involved in digitalisation and virtualisation include

- digitalisation
- cloud computing
- data
- mobile computing
- social media.

Oesterreich and Teuteberg (2016) found that simulation tools/simulation models, BIM, automation, augmented/virtual/mixed reality and RFID were top ranked in relevant scientific publications. IoT/IoS, robotics, augmented/virtual/mixed reality and BIM were top ranked in relevant practical publications.

8.2. Benefits of smart construction to the industry

The economic and social benefits of technology application cannot be over-emphasised. Studies have identified productivity improvements, time and cost savings (Ebekozien and Samsurijan, 2022; Shan *et al.*, 2012), improved construction processes and product visualisation, virtual team's globalisation, stakeholders' value prepositions, information exchange, competitive advantage, increased client satisfaction and quality enhancement (Eastman *et al.*, 2011). Other benefits of the digitalisation of construction are document classification, human resources, information automation, project data visualisation (Chiu and Russell, 2013) and labour productivity (Poirier *et al.*, 2015). The mechanism uses/modifies changing environmental conditions to monitor the behaviour of construction workers. Some of the benefits of smart construction projects have been identified as enhanced vision and strength, enhanced communication, better health and safety, sustainability, enhanced human capabilities, improved productivity and work efficiency, and automation (Al Qady and Kandil, 2014; Chen *et al.*, 2022; Edirisinghe, 2019). In addition, Ebekozien and Samsurijan (2022) found integrated project delivery, project planning and administration, service delivery, improved smart and sustainable projects, improved achievement of clients' objectives and the success of construction projects as the advantages of smart construction. Oesterreich and Teuteberg (2016) summarised the benefits of Industry 4.0 for construction projects as improved sustainability, improved image of the industry, enhanced construction site safety, improve customer relationships with other parties (Vorakulpipat *et al.*, 2010), improved communication and collaboration, improved quality of construction projects (Brandon and Kocaturk, 2009), enhancements in on-time and on-budget integrated project delivery and construction cost and time savings.

Ebekozien *et al.* (2023a) found that the risks of building collapse can be mitigated or prevented using smart construction. As relevant digital technologies for management building collapse risk, they identified

- ultra-wide band (UWB)
- computer vision-based technology
- GPS
- RFID
- video surveillance systems
- virtual reality systems
- IoT/IoS
- construction site safety equipment (e.g. Daqri smart helmet)
- physiological status monitoring (PSM).

RFID, GPS and UWB are safety warning systems and sensor-based technologies, which can advance construction safety management by providing real-time accuracy (Ebekozien *et al.*, 2023b). Videos or photos and analysis of two-dimensional (2D) and 3D images are collected via computer vision-based technology. Video surveillance systems are used for monitoring construction environments and improving the health and safety of workers (Ebekozien *et al.*, 2023a). The Daqri smart helmet is a wearable computing device that can be used to display 3D visuals.

8.3. Encumbrances facing smart construction

Despite the numerous benefits of smart construction to the industry, applications are still low. Attempts to incentivise the take-up of digital technology by construction industries in developing countries have been challenging. Ebekozien and Samsurijan (2022) classified the hindrances as the most severe, severe and fairly severe. The most severe were identified as lack of standardisation for measurement, weak institutional frameworks, lack of ICT-driven construction policy, inadequate infrastructure investment (Ebekozien *et al.*, 2023b), extra monetary limitations on clients and developers, lax attitude to innovation by governments and lack of investment in research. Ebekozien *et al.* (2023c) identified 22 perceived hindrances facing blockchain technology (a component of fourth industrial revolution (4IR) or Industry 4.0) applications in Nigeria and clustered them into three – those related to employees, employers and government. Among the hindrances that cut across the groups were unclear benefits and gains to the parties involved, insufficient investment in research and development, resistance to change, weak stakeholder satisfaction and low awareness of blockchain technology.

Ebekozien *et al.* (2023a) identified initial implementation costs, hesitation to implement, lax government attitude, absence of savings awareness and higher IT requirements as major encumbrances facing the relevance of digital technologies for the risk of building collapse. Oesterreich and Teuteberg (2016) highlighted 12 major challenges facing Industry 4.0 implementation, including legal and contractual uncertainty, lax regulatory compliance, weak communication networks, absence of data security and protection, higher computing equipment requirements, inadequate standards and reference architectures. Other factors are poor acceptance, lax knowledge management, inadequate expertise, organisational and process change, high implementation costs and hesitation to adopt new technology. These challenges are like the encumbrances facing smart construction in developing countries, and they are significant issues influencing technological uptake.

8.3.1 Economic issues

Economic issues are a major factor in the poor acceptance of smart construction by users and contractors. Ebekozien and Samsurijan (2022) and Ebekozien et al. (2023d) noted that the implementation cost is a major hindrance to taking digital technologies beyond experiments. Exoskeletons remain in their infancy despite promising benefits, especially in developing countries because of inadequate affordability.

8.3.2 Standardisation of technologies

Standardisation of technologies is a trending issue in the industry. Scholars (Ebekozien and Samsurijan, 2022; Ebekozien et al., 2023c; Erdogan et al., 2010) have highlighted the need for standardisation and the adoption and understanding of policies and regulations regarding construction-related technologies. Ebekozien and Samsurijan (2022) found that the absence of standards limits the widespread adoption of technology adoption in construction projects in Nigeria.

8.3.3 Technology diffusion

Technology diffusion is about innovation diffusion theory, which measures people's perceptions of innovation attributes (Rogers, 1995), including trialability, observability, complexity, compatibility and relative advantage.

- Trialability is the level to which novelty can be experimented, which includes site testing and tests before real use.
- Observability is the level to which the outcomes of the innovation are visible.
- Complexity is the level of understanding for use of the technology.
- Compatibility is the ability of the technology to coexist
- Relative advantage is the degree of benefit to a company.

The theory emphasises that these factors influence technology adoption, as evidenced by the absence of mature technologies. This includes RFIDs and gaps in other emerging areas such as augmented reality (AR), BIM and safety. The attitude to change is an organisational challenge that influences technology adoption. Besides these factors, environmental factors such as the location of the business, the government and competitors are issues that can hinder technology adoption. Government-driven incentives can change the game plan of stakeholders regarding adopting new technology, as practices in the UK and Singapore have shown regarding BIM adoption (Ebekozien and Samsurijan, 2022).

8.3.4 Technology acceptance

The degree of acceptance and subsequent adoption of new technology is important, irrespective of its reliability and generalisation. It is thus important to engage end-users in the development process, from the start to the finish of the innovation, including test running on-site to ensure users' requirements and expectations have been addressed. Erdogan *et al.* (2010) and Ebekozien and Samsurijan (2022) argued that 'user-centred IT' should be locally developed. This is still an issue in many developing countries because of unfulfilled users' requirements. The absence of industry practitioners to enable researchers/inventors to capture their requirements at the early stage of system development may have contributed to the poor adoption.

8.3.5 Technology limitations

Communication technologies, software and hardware are the future's main technological components of smart construction. Regarding smart construction, hardware and sensors are integral parts. However, the limited battery life of sensors and devices, low computational power and low reliability are challenging issues facing the adoption of technology, especially in developing countries with erratic power supplies. The sensitivity of sensors and data accuracy can influence the functionality of the system. For example, limited hardware capacity has been identified as a challenge to the adoption of AR applications (Meza *et al.*, 2014). Besides these factors regarding validation, scalability (the ability of the system to grow/modify) and generalisability are critical issues that should be considered. A pass mark validation for generalisability implies that the device can be used on other sites, so technology inventors/researchers should ensure that the new technology is applicable and generalisable on any site.

8.4. The future of smart construction

The future of the industry is evolving as more stakeholders are embracing technologies. However, digital technology usage is slow compared with other sectors such as manufacturing and banking (Edirisinghe *et al.*, 2014; Hosseini *et al.*, 2013), especially in developing countries (Ebekozien and Aigbavboa, 2021; Ebekozien and Samsurijan, 2022; Ebekozien *et al.*, 2023a, 2023c, 2023d). Xu *et al.* (2014) and Ebekozien and Samsurijan (2022) noted that the construction sector's adoption challenges are like those in other sectors. These include reluctance or refusal to transform for technology acceptance and people's attitudes. However, the construction industry's unique attributes attract additional hindrances to implementing technology to its full potential. These attributes include the heterogeneity of construction sites and the different backgrounds of the stakeholders involved (Bowden *et al.*, 2006; Navon and Sacks, 2007). The construction industry's future pervasive computing paradigm comprises seamlessly networked computational elements and sensors, supported by entrenched intelligence and innovative digital technologies (Edirisinghe, 2019). These applications should be fixed to various objects that form a network of connections with other devices, with these tools connected to the internet. Thus, this section is divided into three sub-themes – future construction workers, future construction supply chains and future smart construction sites.

- It is envisioned that the workers on construction sites in the future will wear smart wearables such as smart safety glasses (BIMPlus, 2017), smart hard hats (Zhang *et al.*, 2015) and smart e-textiles. Construction workers' smart glasses will visualise data and retrieve real-time data, including automatic BIM (Yeh *et al.*, 2012). Smart glasses will be embedded with data and workers can view refresher safety training linked with BIM on the screen. Edirisinghe and Blismas (2015) and Dalux (2017) reported that smart glasses may be integrated into helmets with the support of AR-based smartphone apps. Workers will need GPS to access a smart hard hat (Zhang et al., 2015). The system will automatically analyse and update workers' productivity (Cheng et al., 2013). Regarding e-textiles, sensors could be embedded in 'smart clothes' to monitor the behaviour of workers.

- The future construction site supply chain (construction procurement and project management) will involve a productive, robust, embedded intelligence. Smart technologies will be used to digitalise as-built models (GeoSLAM, 2017).
- Smart construction sites will be embedded with intelligence regarding safety management. This is germane for future smart construction. Safety management systems will have various objectives. Personal protective equipment (PPE) will be embedded with sensor technologies and supported with on-site sensors. Regarding site plant and equipment operation, the focus of safety management is on on-site construction. This is because plant and equipment have the potential to cause serious hazards. On-site plant will carry tags and be integrated with the support of a fixed tower crane operation or BIM-based tower (Lee et al., 2012). One benefit of this is that workers will be able to view the plant's safe operating procedures screen by scanning a barcode (Lingard et al., 2015). For an effective real-time health and safety management system, refresher training based on legislative requirements for safe work procedures will be automated to mitigate site accidents. Future smart construction sites will be proactive against accident risk through automated controls for collision avoidance and accidents. Also, the environment conditions (air pollution, ultraviolet light, humidity and temperature) of construction sites will be checked and visualised with the support of BIM (Riaz et al., 2014).

More research is required to contribute to the context of 'future pervasive smart construction'. Underinvestment in research and development (local content) in many developing countries, hindering the industry's innovation and technological progress, has not helped the capabilities of new technologies. In the process of technology development, benchmarking new technologies guides development, standardisation and so on. There are research gaps to fill, especially in developing countries, because of the lax embracement of new technologies. Stakeholders in the construction industry are perceived to be afraid of new innovative technologies. This narrative needs to change to open up opportunities for smart construction contributions to the paradigm shift. Regarding development, implementation and technology limitations, technology developers should address these issues through research with the support of the government and other key stakeholders. Governments should lead with policies and programmes encouraging construction companies and other stakeholders to embrace construction digitalisation by investing in it. This should include research centres where construction workers and other practitioners can reskill and upskill regarding new smart construction technology and applications.

It is argued that smart construction will come to realisation soon because it has far-reaching implications for construction industry stakeholders. Besides the economic advantages that come from improved collaboration, quality, efficiency and productivity, smart adoption could improve the industry's image and enhance sustainability and safety in the long run. Manifold technology limitations, acceptance, diffusion, social, technological standardisation, environmental and economic encumbrances should all be embraced. This would contribute to construction economics by offering a comprehensible definition, a discussion of smart building and its relationship with the construction industry.

As the issues facing smart construction are gradually being addressed, it is suggested that governments – through collaboration with regulatory ministries/departments/agencies, professional regulatory bodies (such as the Architects Registration Council of Nigeria (ARCON), the Quantity Surveyors Registration Board of Nigeria (QSRBN) and the Council of Registered Engineering in Nigeria (COREN)), contractors and researchers in the construction domain – try to bring these auspicious technologies to the built environment sector in developing countries. The need for developing countries' governments to study and modify strategies in other countries (e.g. in Singapore programmes such as SkillsFuture, Smart Country Scholarships and Fellowships and TechSkills Accelerator), supported with an enabling environment, cannot be over-emphasised. The government in Singapore has developed an enabling digital environment with sustainable research and skills development investment for developing policies and programmes for a smart country.

8.5. Summary

This chapter covered the concept and characteristics of smart construction. Smart construction integrates construction digitalisation processes to improve productivity and output. The benefits of smart construction to the industry were discussed, with a focus on developing countries. The encumbrances facing smart construction applications were identified and clustered into technology limitations, acceptance, diffusion, technologies standardisation and economic encumbrances. The future of smart construction in a developing country scenario was explored and measures to improve the applications of smart construction technologies in the 21st century were suggested.

REFERENCES

Al Qady M and Kandil A (2014) Automatic clustering of construction project documents based on textual similarity. *Automation in Construction* **42**: 36–49.

BIMPlus (2017) Internet-enabled gloves to improve site safety unveiled. See www.bimplus.co.uk/technology/internet-enabl7ed-glo9ves-impr8ove-site-safety/ (accessed 31/10/2023).

Bowden S, Dorr A, Thorpe T and Anumba C (2006) Mobile ICT support for construction process improvement. *Automation in Construction* **15(5)**: 664–676.

Brandon PS and Kocaturk T (2009) *Virtual Futures for Design, Construction and Procurement.* Blackwell Publishing, Oxford, UK.

Carbonari A, Giretti A and Naticchia B (2011) A proactive system for real-time safety management in construction sites. *Automation in Construction* **20(6)**: 686–698.

Chen X, Chang-Richards AY, Pelosi A *et al.* (2022) Implementation of technologies in the construction industry: a systematic review. *Engineering, Construction and Architectural Management* **29(8)**: 3181–3209, https://doi.org/10.1108/ECAM-02-2021-0172.

Cheng T, Teizer J, Migliaccio GC and Gatti UC (2013) Automated task-level activity analysis through fusion of real time location sensors and worker's thoracic posture data. *Automation in Construction* **29**: 24–39.

Chiu CY and Russell AD (2013) Design of a construction management data visualisation environment: a bottom-up approach. *Automation in Construction* **35**: 353–373.

Dalux (2017) *Dalux TwinBIM*. See https://www.dalux.com/dalux-field/twinbim/ (accessed 16/04/2024)

Dester WS and Blockley DI (1995) Safety—behaviour and culture in construction. *Engineering, Construction and Architectural Management* **2(1)**: 17–26, https://doi.org/10.1108/eb021000.

Dey AK (2001) Understanding and using context. *Personal and Ubiquitous Computing* **5(1)**: 4–7.

Eastman C, Teicholz P, Sacks R and Liston K (2011) *BIM Handbook: A Guide to Building Information Modelling for Owners, Managers, Designers, Engineers and Contractors*. Wiley, Hoboken, NJ, USA.

Ebekozien A and Aigbavboa C (2021) COVID-19 recovery for the Nigerian construction sites: the role of the fourth industrial revolution technologies. *Sustainable Cities and Society* **69**: 102803, https://doi.org/10.1016/j.scs.2021.102803.

Ebekozien A and Samsurijan MS (2022) Incentivisation of digital technology takers in the construction industry. *Engineering, Construction and Architectural Management* **31(4)**: 1373–1390, https://doi.org/10.1108/ECAM-02-2022-0101.

Ebekozien A, Thwala WD, Aigbavboa C and Samsurijan MS (2023a) Investigating the role of digitalisation in building collapse: stakeholders' perspective from unexplored approach. *Engineering, Construction and Architectural Management* **31(13)**: 23–40, https://doi.org/10.1108/ECAM-04-2023-0337.

Ebekozien A, Aigbavboa C, Aliu J, Thwala DW and Emchay EF (2023b) Improving safety on building project sites: the role of sensor-based technology. In *Towards a Sustainable Construction Industry: The Role of Innovation and Digitalisation. Proceedings of 12th Construction Industry Development Board (CIDB) Postgraduate Research Conference* (Aigbavboa C, Thwala W and Aghimien D (eds)). Springer, Cham, Switzerland, pp. 23–32

Ebekozien A, Aigbavboa C and Samsurijan SM (2023c) An appraisal of blockchain technology relevance in the 21st century Nigerian construction industry: perspective from the built environment professionals. *Journal of Global Operations and Strategic Sourcing* **16(1)**: 141–160, https://doi.org/10.1108/JGOSS-01-2022-0005.

Ebekozien A, Aigbavboa C, Samsurijan MS *et al.* (2023d) Smart contract applications in the built environment: how prepared are Nigerian construction stakeholders? *Frontiers of Engineering Management* **11**: 50–61, https://doi.org/10.1007/s42524-023-0275-z.

Edirisinghe R (2019) Digital skin of the construction site: smart sensor technologies towards the future smart construction site. *Engineering, Construction and Architectural Management* **26(2)**: 184–223, https://doi.org/10.1108/ECAM-04-2017-0066.

Edirisinghe R and Blismas N (2015) A prototype of smart clothing for construction work health and safety. In *Proceedings of the CIB W099 International Health and Safety Conference: Benefitting Workers and Society through Inherently Safe(r) Construction, Belfast, Northern Ireland*, pp. 1–9.

Edirisinghe R, Blismas N, Lingard H and Wakefield R (2014) Would the time-delay of safety data matter? Real-time Active Safety System (RASS) for construction industry. In *Proceedings of the CIB W099 International Conference on Achieving Sustainable Construction Health and Safety, Lund, Sweden*, pp. 564–574.

Erdogan B, Abbott C and Aouad G (2010) Construction in year 2030: developing an information technology vision. *Philosophical Transactions of the Royal Society of London A: Mathematical, Physical and Engineering Sciences* **368(1924)**: 3551–3565.

Fang W, Ding L, Luo H and Love PE (2018) Falls from heights: a computer vision-based approach for safety harness detection. *Automation in Construction* **91**: 53–61, https://doi.org/10.1016/j.autcon.2018.02.018.

GeoSLAM (2017) *Whitepaper. Digital Engineering: Building Reality.* GeoSLAM, Bingham, UK. See https://www.landmark.com.gr/wp-content/uploads/2017/07/Geoslam-Whitepaper-Digital-Engineering-Low.pdf (accessed 16/04/2024).

Gubbi J, Buyya R, Marusic S and Palaniswami M (2013) Internet of things (IoT): a vision, architectural elements, and future directions. *Future Generation Computer Systems* **29(7)**: 1645–1660.

Hosseini MR, Chileshe N, Zuo J and Baroudi B (2013) Approaches of implementing ICT technologies within the construction industry. *Australasian Journal of Construction Economics and Building Conference Series* **1(2)**: 1–12.

Jiang W, Ding Z and Zhou C (2021) Cyber physical system for safety management in smart construction site. *Engineering, Construction and Architectural Management* **28(3)**: 788–808, https://doi.org/10.1108/ECAM-10-2019-0578.

Kim H, Kim K and Kim H (2015) Vision-based object-centric safety assessment using fuzzy inference: monitoring struck-by accidents with moving objects. *Journal of Computing in Civil Engineering* **30(4)**: 04015075, https://doi.org/10.1061/(ASCE)CP.1943-5487.0000562.

Kochovski P and Vlado S (2018) Supporting smart construction with dependable edge computing infrastructures and applications. *Automation in Construction* **85**: 182–192, https://doi.org/10.1016/j.autcon.2017.10.008.

Lee G, Cho J, Ham S *et al.* (2012) A BIM-and sensor-based tower crane navigation system for blind lifts. *Automation in Construction* **26**: 1–10, https://doi.org/10.1016/j.autcon.2012.05.002.

Li RYM (2017) Smart construction safety in road repairing works. *Procedia Computer Science* **111**: 301–307, https://doi.org/10.1016/j.procs.2017.06.027.

Lingard H, Pink S, Harley J and Edirisinghe R (2015) Looking and learning: using participatory video to improve health and safety in the construction industry. *Construction Management and Economics* **33(9)**: 1–12.

Malik S, Fatima F, Imran A *et al.* (2019) Improved project control for sustainable development of construction sector to reduce environment risks. *Journal of Cleaner Production* **240**: 118214, https://doi.org/10.1016/j.jclepro.2019.118214.

Meng Q, Zhang Y, Li Z *et al.* (2020) A review of integrated applications of BIM and related technologies in whole building life cycle. *Engineering, Construction and Architectural Management* **27(8)**: 1647–1677, https://doi.org/10.1108/ECAM-09-2019-0511.

Meza S, Turk Z and Dolenc M (2014) Component based engineering of a mobile BIM-based augmented reality system. *Automation in Construction* **42**: 1–12.

Mihindu S and Arayici Y (2008) Digital construction through BIM systems will drive the re-engineering of construction business practices. In *2008 Conference on Visualisation*. IEEE, Piscataway, NJ, USA, pp. 29–34, https://doi.org/10.1109/VIS.2008.22.

Mills F (2016) BIM and social media. In *Construction Manager's BIM Handbook* (Eynon J (ed.)). Wiley, Chichester, UK, pp. 127–132.

Navon R and Sacks R (2007) Assessing research issues in automated project performance control (APPC). *Automation in Construction* **16(6)**: 474–484.

Niu Y, Lu W, Liu D, Chen K and Xue F (2017) A smart construction object (SCO)-enabled proactive data management system for construction equipment management. In *Computing in Civil Engineering*. ASCE, Reston, VA, USA, pp. 130–138, https://doi.org/10.1061/9780784480830.017.

Oesterreich TD and Teuteberg F (2016) Understanding the implications of digitisation and automation in the context of industry 4.0: a triangulation approach and elements of a research agenda for the construction industry. *Computers in Industry* **83**: 121–139.

Poirier EA, Staub-French S and Forgues D (2015) Measuring the impact of BIM on labour productivity in a small specialty contracting enterprise through action-research. *Automation in Construction* **58**: 74–84.

Riaz Z, Arslan M, Kiani AK and Azhar S (2014) CoSMoS: a BIM and wireless sensor based integrated solution for worker safety in confined spaces. *Automation in Construction* **45**: 96–106.

Rogers EM (1995) *Diffusion of Innovations*, 4th edn. Free Press, New York, NY, USA.

SCOSCC (Safety Committee Office of the State Council of China) (2018) *Production Safety Accidents Occurred in the National Construction Industry in China*. See http://m.people.cn/n4/2018/0725/c203.

Shan Y, Goodrum P, Haas C and Caldas C (2012) Assessing productivity improvement of quick connection systems in the steel construction industry using building information modelling (BIM). In *Construction Research Congress 2012: Construction Challenges in a Flat World*. ASCE, Reston, VA, USA, pp. 1135–1144.

Sutrisna M and Kumaraswamy MM (2015) Cyber-physical systems integration of building information models and the physical construction. *Engineering, Construction and Architectural Management* **22(5)**: 516–535, https://doi.org/10.1108/ECAM-07-2015-0120/full/html.

Vorakulpipat C, Rezgui Y and Hopfe CJ (2010) Value creating construction virtual teams: a case study in the construction sector. *Automation in Construction* **19(2)**: 142–147.

Weiser M, Gold R and Brown JS (1999) The origins of ubiquitous computing research at PARC in the late 1980s. *IBM Systems Journal* **38(4)**: 693–696.

Woodhead R, Stephenson P and Morrey D (2018) Digital construction: from point solutions to IoT ecosystem. *Automation in Construction* **93**: 35–46, https://doi.org/10.1016/j.autcon.2018.05.004.

Xu H, Feng J and Li S (2014) Users-orientated evaluation of building information model in the Chinese construction industry. *Automation in Construction* **39**: 32–46.

Yeh KC, Tsai MH and Kang SC (2012) The iHelmet: an AR-enhanced wearable display for BIM information. In *Mobile and Pervasive Computing in Construction* (Anumba CJ and Wang X (eds)). Wiley-Blackwell, Hoboken, NJ, USA, pp. 149–168.

Zhang S, Teizer J, Pradhananga N and Eastman CM (2015) Workforce location tracking to model, visualise and analyse workspace requirements in building information models for construction safety planning. *Automation in Construction* **60**: 74–86.

Zhou H, Wang H and Zeng W (2018) Smart construction site in mega construction projects: a case study on island tunnelling project of Hong Kong-Zhuhai-Macao Bridge. *Frontier Engineering Management* **5(1)**: 78–87, https://doi.org/10.15302/J-FEM-2018075.

Index

4IR (fourth industrial revolution) *see also* Industry 4.0 106

accounting 11, 34, 62
advertisement 33, 48
AI (artificial intelligence) 103
analytical estimating 20–21
appropriation 33, 48
approximate estimating 10, 12, 15–21, 25, 32, 39
approximate quantities 6, 20, 32, 38
AR (augmented reality) 107, 108, 109
architect 5, 6–7, 27–8, 29–32, 35–6, 40, 41
ARCON (Architects Registration Council of Nigeria) 110
ARR (average rate of return) 62–4

BCIS (UK Building Cost Information Service) 37, 38, 39, 42
   building cost indices 68, 71
   cost analysis system 41, 42
   tender price index 71
BIM (building information modelling) 104, 105, 107–10
blockchain technology 106
BOQ (bill of quantities) 5,6, 8, 20, 32, 36, 39
budgeting 11, 67
buildability 23
building
   contractor 28, 29, 32, 71, 84
   cost 20, 24, 52, 53, 67, 73, 80, 82, 86–8
   cost indices 68–71
   design costs 21
   economics 10–12, 15, 98
   elements 30,38, 40, 41, 48, 51
   estimate 12
   green 55, 91–2, 96–8

green certification 96
grouping 23
height 16, 17, 19, 23
indicators 71
information modelling (BIM) 104, 105, 107–10
investment 95
shape 17, 18, 22–3, *23*
size 21, 22, *22*, 75
built environment sector *see* construction industry

cash flow 7, 11, 55–64, 75, 77, 78
choice 10, 11, 15, 21, 29, 73
choosing cost indices 72–3
client 7–8, 27
   government as 9–10
climate change 93, 96
cloud computing 104, 105
cobra effect 91
comparative cost planning 38–9
concise cost analysis 42
construction
   data 67, 68, 69
   developments 24–5
   digitalisation of 105, 109
   economics 4, 10–12, 15, 98, 110
   funding 97
   future smart sites 108, 109
   future supply chains 108,109
   future workers 108, 109
   green 55, 91, 98
   infrastructure 91
   maintenance-free 24
   management 3, 36
   market 3, 9

methods 52, 91, 94, 96
policy 106
practices 92–4, 96–8
practitioners 10, 36, 51
sector 1–4, 8–9, 94, 96, 103, 108
site productivity 11
site safety 105
smart *see* smart construction
sustainable 91–8
sustainable materials 12
technology 4
technology 4, 109
workers 4, 8, 9, 92, 104, 105, 108, 109
construction industry 37, 91, 94, 95, 108, 110
   accidents 103
   approximate estimating 15–21
   characteristics 2–3
   cost indices 67–73
   definition 1, 2
   design variables 21–5
   developing countries 1, 2, 67
   developing countries 67
   economic regulator 8–9
   economics 1–12, 15, 98, 110
   Ghana 1, 2
   government action on 9–10
   influences on activity 2
   job creation 2
   Nigeria 1, 2, 52, 98
   stakeholders 4–8, 5, 53, 97, 104, 109
   subsectors 3
constructional details 24
context awareness 104
continuous training 97–8
contract 5–8, 72
   administration 27–36, 48
   cancellation 8
   documents 15
   management 52, 53
   period 24, 82
   price 39, 69
contractors 8, 34–6
   costs 6, 69
   method (property valuation) 78
   Nigeria 69, 92, 110

subcontractors 4, 5, 8, 9, 28, 30, 32, 36, 39
sustainability and 97
controlling 7, 10–12, 27, 31, 39, 40, 72, 76
COREN (Council of Registered Engineering in Nigeria) 110
cost
   analysis 18, 20, 27, *29*, 38–48, 52–3, 67, 70
   assessment 12, 72
   checking 28, *29*, 31, 32, 37, 38, 41
   contractor's 6, 69
   control 8, 12, 27, *29*, 31, 37, 40, 41, 48, 98
   controlling 12, 31, 40
   data 15, 19, 39, 41, 48, 67, 69, 71
   Davis Langdon forecasts 71
   development 37, 78, 86
   economic 98
   estimation 6, 31, 39, 51–3, 55, 66, 73
   expert 21
   finance 82, 86
   implication *22*, 23, 30, 31, 38, 40, 52, 92
   indices 67–73
   information 38, 39, 69
   information management 39
   management 4
   manager 4, 5, 38
   method (property valuation) 75
   models 21, 39, 54
   opportunity, of 11
   plan 28, *29*, 30, 31, 37–8, 40, 41, 47
   planning 19, 20, 27, 30–32, 36–41, 47–8, 69, 98
   planning techniques 38–9
   presentation 37, 38, 39
   project 7, 20, 24–5, 31, 36–7, 40, 41, 51–2, 67, 69
   R's, the 24
   terminology 40–41
cube method 16–17

data
   big 104
   collection 54, 103
   construction 67, 68, 69
   cost 15, 19, 39, 41, 48, 67, 69, 71
   databases 19, 69, 71

Index

management 104
   quality 69
   real-time 108
   unreliable 54, 77
David Langdon cost forecasts 71
design 3
   detailed 6, 15, 28, 31–2, 37, 41
   economic 19, 23
   project 11, 42
   scheme 29, 30–31, 37
   team 4–7, 21, 27–38, 40, 41, 48, 51, 69
   variables 12, 15, 21–5
detailed cost analysis 42
detailed design 6, 15, 28, 31–2, 37, 41
determining property value 75
developer's budget 75, 77, 79–88
developer's profit 79, 83–8
developing countries/nations 1, 2, 8–9, 40, 51–3, 69–73, 75, 77, 95–6, 106–10
digitalisation and visualisation 104, 105

economic
   analysis 94
   costs 98
   design 19, 23
   development 1, 2, 71, 72
   downturns 2, 3
   encumbrances 94, 95, 110
   environment 95
   forecasts 10, 11
   indicators 69, 70
   issues 10, 94, 98, 107
   life/lives 58, 62, 64
   macro 94–5
   measures 94
   meso 95
   micro 95
   regulator 8, 12
   sustainability 94
economics 23
   construction industry and 1–12
   construction/building 4, 10–12, 15, 98, 110
   smart construction 103–10
   study of 11
   sustainable construction 88, 91–8

electrical engineers 6
elemental cost planning 38, 39
elemental estimating 20
end users 1, 3, 51, 92, 93, 94, 108
engineers 5, 7, 15, 27–8, 30, 32
environmental sustainability 52, 93, 98
ERR (external rate of return) 64
evaluating 11, 51, 72, 94
explicit time-variable hedonic index 72

factor cost indices 68, 70–71
feasibility report 30
financial methods 21
floor area method 18–19
FMW (Federal Ministry of Work) 35
forecasting 10, 11, 21, 39, 40, 67, 68, 69

GDP (gross domestic product) 1, 11, 69, 70, 75
GDV (gross development value) 78, 79–81, 83–8
governance 95, 97
government
   action 9–10, 12
   agency 4, 7, 79, 80
   client, as 9, 10
   intervention 94, 97
   legislation 9, 24, 54, 75, 77
   policy 9, 10, 76
   regulations 9, 97
GPS (Global Positioning Systems) 104, 106, 109
green building life-cycle costing 55
green policies 92, 96
greenhouse gases 91

health and safety 92, 105, 106, 109
hedonic price modelling (property valuation) 79
hedonic regression method 72
HERR (historical external rate of return) 64–5
housing and construction statistics 71

ICT (information and communication technology) 103, 104, 106
improved traditional life-cycle costing 55
inception stage 27, 29, 37, 40, 51–2
income method (property valuation) 75

Industry 4.0 103, 104, 105, 106, 107
information management 4, 39
innovation diffusion theory 107
investment appraisal 55–65, 76, 77, 88
investment method (property valuation) 78–9
IoS (Internet of Senses) 104, 105, 106
IoT (Internet of Things) 104, 105, 106
IRR (internal rate of return) 58, 59, 60, 64
IUA (International Union of Architects) 6

LDC (less developed countries) 94, 95
legal fees 82, 84
life-cycle costing (LCC) 10, 24, 41, 51–65, 94, 96
   analysis 52–3
   conventional methods 55, 61–5
   discounting methods 55–61
   encumbrances 53–4
   fundamentals 51–2
   green building 55
   improved traditional 55
   methods 55
   non-traditional 55
   tools/techniques 55–65, 68
   traditional 55
   worked examples 56–65

macroeconomic 11
   encumbrances 94–5, 98
MARR (minimum attractive rate of return) 56, 57, 58
mechanical engineers 6
mesoeconomic encumbrances 95, 98
microeconomic encumbrances 95, 98
mobile computing 103, 105
model 21, 97, 104
   BIM *see* BIM (building information modelling)
   cost 21
   explicit time-variable 72
   green certification of buildings 96
   hedonic price 79
   LCC parameters 55
   property development process 76, 77
   risk 77
   strictly cross-sectional 72
   three pillars 93–4, 96

Nationwide Building Society index 71–2
necessity postponability 61
NEDO (National Economic Development Office) method 72
NFV (net future value) 56–8, 64, 65
NIC (newly industrialised country) 94, 95
Nigeria 9, 69
   appropriation 33
   ARCON (Architects Registration Council of Nigeria) 110
   construction industry 1, 2, 52, 98
   contractors Nigeria 69, 92, 110
   COREN (Council of Registered Engineering in Nigeria) 110
   cost indices 67–73
   cost information 39
   digitalisation 12
   funding issues 7
   government 9, 24, 96
   life-cycle costing (LCC) 51–3, 56
   NIQS (Nigerian Institute of Quantity Surveyors) 4, 5, 39, 72
   property development 76, 77
   property investment 75–7
   property market 76
   property value 75–88
   QSRBN (Quantity Surveyors Registration Board of Nigeria) 110
   quantity surveying programmes 10
   smart construction 106
   sustainable development 91–3, 95–7
non-collaborative work practices 1
non-traditional life-cycle costing 55
NPV (net present value) 55–60

optimal investment criterion 64
outline proposal 27, *29*, 30, 37

payback method 61–2
perimeter method 19
pervasive computing 103, 108
plan of work 27–8, 30, 31, 48
plan shape 22–3
planning efficiency 22
plant use 25
PPE (personal protective equipment) 109

Index

practical completion 7, 28, 35–6, 51
prequalification 33–4
probable costing *see* approximate estimating
production information 28, *29*, 32, 37
professional fees 79, 82, 84, 85, 86, 87
profit or account method (property
        valuation) 78
project
  constellations 7
  costs 7, 20, 24–5, 31, 36–7, 40, 41, 51–2,
    67, 69
  delivery 4, 8, 104, 105, 106
  design 11, 42
  management 2, 104, 109
  performance 1, 4, 7, 28
  planning 28, *29*, 35, 48, 105
  unit rates 16
property
  development 76, 77
  income 10
  investment 75–7
  market 76, 80
  valuation 73, 75–9
  valuation methods 77–9
  value 75
property development process model 76, 77
public sector borrowing 8

QS (quantity surveyor) 4–6, 15–17, 19–21, 24,
    27–32, 35–40, 68, 71, 72
QSRBN (Quantity Surveyors Registration Board
    of Nigeria) 110

R's, the (running, repairs, replacement costs) 24
reinstatement method (property valuation) 79
residual method (property valuation) 78
Revised Public Procurement Act (2007) 32
RFID (radio frequency identification) 104, 105,
    106, 107
RIBA (Royal Institute of British Architects) 16,
    37
  plan of work 27–8, 30, 31, 48

safety management 106, 109
sales comparison method (property valuation)
    75, 78

scarcity 11, 19, 25
scheme design 29, 30–31, 37
sensors 103, 108, 109
simulation and modelling 104, 105
Singapore 1, 2, 69, 107, 110
site
  clearance 24
  condition 16, 24–5, 27
  considerations 21
  finance 84, 86
  layout 28, 35, 84
  location 21, 24, 79
  operations 35
smart construction 75, 98
  advantages 105
  benefits of 105–6
  economic issues 107
  economics of 103–10
  encumbrances 106–8
  foundations of 103–5
  future of 108–10
  standardisation 107
  studies 104
  technology acceptance 108
  technology diffusion 107
  technology limitations 108
smart factories 104–5
smart technologies 103, 109
social sustainability 93–4
stakeholders 4–8, *5*
storey height 17, 18, 19, 23, 42
storey-enclosure method 19
strictly cross-sectional hedonic index 72
structural engineers 7
subcontractors 4, 5, 8, 9, 28, 30, 32, 36, 39
superficial method 18–19
sustainability 40, 52, 53, 92, 96, 105, 109
  cost of 98
  economic 94
  environmental 93
  social 93–4
sustainable
  architecture 91
  building code 96, 7
  construction *see* sustainable construction
  development 55, 91, 93, 94, 95

funding 97
materials 54, 91
practices 92, 97, 98
research 110
sustainable construction 91–8
   benefits 92–3
   encumbrances (developing countries) 95–6
   encumbrances (general) 94–5
   principles 93–4
   principles 93–4
   recommendations 96–8
   three pillars model 93–4, 96

technological advancements 3
technology
   acceptance 108, 110
   adoption 107, 108, 110
   benefits 105
   blockchain 106
   computer vision-based 106
   construction 4, 109
   development 109
   diffusion 107, 110
   digital 106, 108
   limitations 108, 110
   smart 109
   standardisation 107, 110
tender
   action 28, *29*, 32–5
   documents 28, 32, 34
   invitation 34
   opening 34
   price indices 68–70, 71
   stage 32, 37, 48
three pillars model 93–4, 96
traditional life-cycle costing 55

unit method 15–16, 81
United Nations Sustainable Development Goals 91
usable/non-usable areas 22, 23
user
   end 1, 3, 51, 92, 93, 94, 108
   requirements 21, 29, 108
UWB (ultra-wide band) 104, 106

value for money 4, 10, 11, 38, 40, 51

World Commission on Environment and Development 91

Printed in the USA
CPSIA information can be obtained
at www.ICGtesting.com
JSHW011420030924
69236JS00012B/235